Fishing Wars and Environmental Change in Late Imperial and Modern China

Harvard East Asian Monograph 325

Fishing Wars and Environmental Change in Late Imperial and Modern China

Micah S. Muscolino

Published by the Harvard University Asia Center
Distributed by Harvard University Press
Cambridge (Massachusetts) and London 2009

Printed in the United States of America

The Harvard University Asia Center publishes a monograph series and, in coordination with the Fairbank Center for Chinese Studies, the Korea Institute, the Reischauer Institute of Japanese Studies, and other faculties and institutes, administers research projects designed to further scholarly understanding of China, Japan, Vietnam, Korea, and other Asian countries. The Center also sponsors projects addressing multidisciplinary and regional issues in Asia.

Library of Congress Cataloging-in-Publication Data
Muscolino, Micah S., 1977–
    Fishing wars and environmental change in late imperial and modern China /
Micah S. Muscolino.
        p. cm. -- (Harvard East Asian monograph ; 325)
    Includes bibliographical references and index.
    ISBN 978-0-674-03598-0 (alk. paper)
        1. Fisheries--China--History. 2. Fisheries--Environmental aspects--China--History. 3.
Fisheries--Political aspects--China--History. 4. China--History, Military. I. Title.
    SH297.M87 2009
    338.3'7270951--dc22

                                                                    2009022547

Index by the author

∞   Printed on acid-free paper

Last figure below indicates year of this printing
18 17 16 15 14 13 12 11 10

Acknowledgments

I owe a tremendous intellectual debt to my graduate advisors: William Kirby, Philip Kuhn, and Andrew Gordon. I have benefited greatly from their rigorous training, expert guidance, and unfailing encouragement.

Many thanks go to the individuals who facilitated my research for this study. Lin Man-hong arranged my stay at Academia Sinica's Institute of Modern History, and the staff of the Shanghai Academy of Social Sciences expedited much of my research in China. Liang Jingming of Zhejiang University was a gracious host during my time in Hangzhou. The search for materials in Zhoushan would have failed without help from Sang Xiuping, director of the Zhoushan Municipal Library. A Fulbright IIE Fellowship and a grant from the Center for Chinese Studies at Taiwan's National Central Library made this research possible. Harvard University's Graduate School of Arts and Sciences lent further financial assistance with a Sheldon Traveling Fellowship and a Graduate Society Dissertation Completion Fellowship.

I am especially grateful for the friends and colleagues who took the time to read and comment on drafts of chapters: Ravi Bhandari, Ning Chang, Evan Dawley, Timothy George, Xiaofei Kang, Ian Miller, James Millward, Myrna Santiago, Howard Spendelow, Julia Strauss, Donald Sutton, William Tsutsui, Brett Walker, and

Gao Yan. Robert Marks, Peter Perdue, and David Pietz read through the project's previous incarnation as a dissertation and offered suggestions for improvement. The two anonymous reviewers for the press pushed me to refine and rethink many of my ideas. Carol Benedict and John McNeill gave detailed and extremely valuable feedback on the revised manuscript. Tommaso Astarita, Alison Games, David Goldfrank, Carl Guarneri, Gretchen Lemke-Santangelo, Amy Leonard, Elizabeth Perry, Aviel Roshwald, Keith Schoppa, Richard Stites, John Tutino, Aparna Vaidik, Timothy Weston, and Wen-hsin Yeh pointed me in the right direction with their advice and encouragement. All errors or deficiencies that remain are my fault alone.

The greatest thanks by far go to my family. This book is dedicated to them.

M.M.

Contents

Abbreviations

The following abbreviations are used in the Notes and Works Cited.

AH	Academia Historica, Taiwan
CMB	*Chongming bao* (Chongming news)
DFZZ	*Dongfang zazhi* (Eastern miscellany)
DHMB	*Dinghai min bao* (Dinghai people's news)
DHZB	*Dinghai Zhou bao* (Dinghai Zhoushan news)
IMH	Institute of Modern History Archives, Academia Sinica, Taiwan
NHA	Number Two Historical Archives, Nanjing
PHC	Nationalist Party Historical Commission Archives, Taiwan
SB	*Shenbao* (Shanghai news)
SCYK	*Shuichan yuekan* (Marine products monthly)
SMA	Shanghai Municipal Archives
ZMA	Zhoushan Municipal Archives
ZPA	Zhejiang Provincial Archives
ZSB	*Zhejiang shang bao* (Zhejiang commercial news)

Fishing Wars and Environmental Change in Late Imperial and Modern China

Introduction

Countless fishing boats, gathered in a harbor.
The flickers of fishing lanterns flow in the rippling waves.
Highest heaven's stars have fallen in the dead of night.
Shining everywhere, like a coral islet in the sea.

—Liu Menglan, "Qugang yudeng"[1]

Written in the 1830s, this poem paints an evocative picture of a nighttime scene in the Zhoushan Archipelago, the chain of islands off the coast of Zhejiang and Jiangsu provinces that makes up China's most important marine fishing ground. It vividly captures the tranquility of fishing boats anchored in the harbor, as the light from their lanterns plays upon the water and illuminates the darkness. This idyllic imagery gives no inkling of the tremendous changes that would occur beneath these waves during the next one and a half centuries.

China currently faces environmental issues ranging from severe water shortages and land degradation to climate pollution and decreasing biodiversity.[2] These challenges extend to the marine environment, where pollution and overfishing have damaged China's fishery resources.[3] The depletion of China's fisheries, like its many other environmental problems, intersects with global ecological transformations. In recent years, the accelerating loss of marine biodiversity has brought the world's fisheries to a state of crisis. This trend has led some researchers to predict a collapse of all marine fish and seafood species by 2048 unless steps are taken to stop overfishing, control pollution, and protect ocean habitats.[4]

The degradation of the earth's marine ecosystems, like all environmental issues, derives from alterations of the natural environment in times past. Coming to grips with contemporary challenges demands a better understanding of how people have historically generated, perceived, and responded to environmental change. With this goal in mind, the present study explores reciprocal interactions between society and environment in the Zhoushan maritime region. The temporal scope extends from the upsurge in human migration to these islands in the nineteenth century to the exhaustion of the region's most commercially important fish species in the 1970s. The history of Zhoushan's fisheries thus illuminates long-term environmental processes spanning the Qing dynasty (1644–1911), the Republican period (1911–49), and the People's Republic of China (1949–).

This history of the Zhoushan maritime region marks the first major study of Chinese fisheries from the perspective of environmental history. With China's vast coastline and extensive system of inland lakes and waterways, fishing has long been an important economic activity. For people in southeast China, especially coastal areas like the Zhoushan Archipelago, fish and other marine products provided dietary protein, as well as sources of income.[5] Previous research has greatly advanced our understanding of the history of human interactions with China's diverse natural landscapes, but interactions with China's marine ecosystems have heretofore remained an untold story. Perhaps because of the enduring assumption that the ocean exists apart from human history, the same almost holds true in the field of environmental history as a whole. To fully comprehend the current state of the world's marine ecosystems, as W. Jeffrey Bolster reminds us, we need to know much more about how long people have been making an impact on the earth's oceans, when warning signs of overexploitation first appeared, and what assumptions and policies made these impacts possible.[6] This history of Zhoushan's fisheries enters these uncharted waters, examining changes in the marine environment off China's eastern coast, along with their social, economic, and political ramifications.

Turning our attention to the oceans highlights ecological processes that transcend the boundaries of the nation-state. Fish are highly mobile and often migrate great distances during their lifespan, moving through waters claimed by multiple political entities. The migratory character of fish makes it impossible to analyze a single fishery in isolation from larger-scale environmental changes. Although focusing specifically on the Zhoushan Archipelago's maritime environment, this study builds on previous literature in the field of Chinese environmental history by analyzing the intersection of local, regional, and transnational ecological trends.

In terms of periodization, most research on China's environmental history has focused on either the Qing or the People's Republic after 1949. Along with trends that emerged during the Qing and PRC, however, developments in the Republican period had a tremendous impact on Zhoushan's marine environment. Historians of Republican China now characterize it as a time of economic growth, expanding connections between China and the world, and the formation of a modern Chinese nation-state. These modern transformations—much like China's recent rise as a global economic power—also left their mark on China's landscapes and seascapes. Bringing the Republican period into the literature on China's environmental history makes it possible to identify ecological processes that linked the late Qing and the PRC. Pressures on Zhoushan's fisheries took shape during the Qing, but developments in Republican China intensified these ecological changes, prefiguring trends that became readily apparent under the People's Republic. The dynamics of environmental change, as the history Zhoushan's fisheries demonstrates, defy conventional political periodization.

The Problem of the Commons

Intensified human exploitation of Zhoushan's marine ecosystem was one of many environmental processes that spanned late imperial and modern China. Beginning in the eighteenth century, commercial development, population growth, and internal migration combined to place unprecedented demands on the natural environment. In many regions, these changes aggravated problems associated with the use of common-pool environmental resources such as irrigation systems, forests, pasturelands, and, of course, fisheries. All types of common resources have two main characteristics. First, it is difficult to exclude individuals from using them. Second, the benefits derived from the resource by one individual subtract from those available to others.[7] The history of Zhoushan's fisheries illustrates how users regulated the use of common resources in one particular local setting over a period of more than two centuries. Their strategies, however imperfect, reveal some of the ways in which people in late imperial and modern China struggled to coordinate the use of natural resources in the face of unprecedented ecological pressures.

Commons problems may assume a variety of forms. Under conditions of open access, when no rules regulate the use of the commons, conflict can be unavoidable and violence routine. Without rules coordinating the use of resources, unrestrained competition threatens to diminish profits by diverting time, energy, and assets into violent disputes. The problem here is not necessarily overexploitation but violent conflict over access to productive resources. Second, as people take from a finite, common pool of environmental resources, they leave less behind for other users. If no rules limit access to common resources, people take as much as they can as quickly as possible to avoid giving them up to others. If the process is unchecked, people end up ruining the resource due to overexploitation.[8]

Fisheries are a prime example of a common resource susceptible to these problems. Since fishers often do not allocate resources efficiently, they fight for productive fishing spots. In these instances, "The problem is not that the resource is being overexploited or that it is on the verge of destruction. Rather, the problem is that some fishing spots are better than others, with conflict erupting over who can fish where."[9] Competition for the best fishing spots and interference between boats, nets, or other gear can lead to intense and costly disputes that cut into benefits obtained from fish stocks. Other problems arise when fishers do not consider the effect that their harvesting activities have on the resource as a whole.[10] Since greater catches by one user have the potential to reduce the amount of fish available to others, everyone takes as much as he can as quickly as he can. More intensive harvesting may begin to decrease the amount of fish left for future use, leading to their eventual exhaustion.[11]

Nevertheless, research in various disciplines has shown that in many instances the users of fisheries and other common resources devise arrangements that regulate how the resources are used, minimizing some (but not necessarily all) of the costs of unrestrained exploitation.[12] These coordinated strategies for solving commons problems appear when people create institutions, or sets of working rules, that limit benefits from the resource to members of a defined group and regulate its exploitation.[13]

The Commons and Chinese Environmental History

The transformation of Zhoushan's marine environment reached unprecedented levels during the eighteenth and nineteenth century, when intensified commercialization and demographic expansion drew China's growing population to move into previously unexploited ecological frontiers. Their migrations involved clearing hillsides and tilling grasslands for agriculture, as well as opening up marine fishing grounds. These socioeconomic trends subjected common-pool resources to greater exploitative pressures, leading to widespread ecological degradation. Furthermore, as market forces and population pressure made natural resources increasingly scarce, people often resorted to violence. Struggles over common-pool resources such as forests, uncultivated lands,

and irrigation systems thus contributed to the endemic violence
that raged in many regions of China during the late Qing and Republican period.[14]

Environmental historians of China have already highlighted
the difficult and precarious task of managing common-pool resources. Most notably, Peter Perdue's and R. Keith Schoppa's local studies of water control during the Qing detail the efforts of
enlightened officials to halt private encroachment on irrigation
systems. Yet even the most far-sighted official interventions had
little hope of success unless they surmounted the barriers to implementation in local society. Commercial and demographic pressure ultimately made it impossible for high-level officials to halt
illegal dike building and land reclamation that paved the way for
future natural disasters.[15]

The obstacles facing official supervision under the Qing were
also apparent in efforts to resolve violent conflicts over natural
resources. Given the limited capacity of overburdened and under-
staffed officials to enforce dispute settlements, they often "directly enlisted the local community to sustain their decisions."[16]
This type of collaboration between officials and local communities fits well with what Philip Huang defines as the "third realm"
between state and society.[17] But historians have paid far more attention to the ingenuity with which the Qing state handled violent disputes over natural resources than to how local communities worked out solutions to these conflicts. What is more, even
less is known about how these institutions fared under the vastly
different political conditions prevailing in the Republican period.

With these issues in mind, this book investigates the question
of how people controlled the Zhoushan Archipelago's fisheries
and to what ends by analyzing the array of private and state interests that shaped struggles for the control of these common-pool
natural resources. What types of institutions did private and state
actors rely on to regulate the use of the fishery? How did relationships between social organizations and the state that emerged
during the Qing change under the different conditions of the Republican period? What types of problems could these arrange-

ments solve, and which ones were they unable to overcome? What does the fate of these institutions tell us about environmental change in late imperial and modern China? Answering these questions, it is hoped, will give us a better understanding of the relationship between ecological changes in China's past and its present environmental challenges.

Overview

Beginning in the mid-Qing, fishermen who followed the seasonal migrations of fish species to the Zhoushan islands from their native regions in Zhejiang and Fujian provinces staked *de facto* proprietary claims to fishing grounds and established rules coordinating their use. Regionally based coalitions divided fishing grounds into sectors, with each group staying in its designated area. Local officials were the final arbiters in many fishing disputes, but they consistently delegated responsibility for implementing and enforcing settlements to the leaders of native-place lodges. Organized by migrants from the same province, county, or town, these regional associations gave fishers and merchants a center for providing essential services and for furthering their common interests. This type of native-place organization was not unique to the Zhoushan region but existed among sojourning merchants and artisans in cities and towns throughout late imperial China, where regional groups claimed different market sectors and sought to regulate trade within them.

Relying on these largely unofficial strategies, those who profited from Zhoushan's fishing grounds averted violent conflicts over the control of common-pool resources. Local religion facilitated definition and enforcement of these rules, enabling fishing communities to cope with the risk and uncertainty that characterized their trade. In the Zhoushan region, native-place ties defined by collective religious observances functioned as an ecological strategy for allocating access to resource space. These local arrangements did not, however, limit economic pressures to intensify exploitation of fish stocks. On the contrary, social institutions were essential for the marketing networks that transformed fish into commodities.

During the Republican period, the modern Chinese state intervened in Zhoushan's fisheries to increase catches and tax revenues by rationalizing exploitation of the marine environment. Rather than alleviating ecological problems, developmental plans aimed at increasing output and maximizing revenues subjected fish stocks to even greater pressures. From the Republican period through the 1950s, the foreign-trained Chinese fishery experts who staffed government agencies argued that rationalized, scientific management could make exploitation of resources more efficient, while preventing their depletion. These assumptions proceeded from the notion that application of scientific and technical expertise could manipulate the environment for maximum production. Whereas social institutions struggled to confront the reality of mounting competition for scarce resources, the modern state's management initiatives were based on a belief that science and technology could make nature endlessly productive.

This vision of development displayed the optimism characteristic of what James Scott terms "high-modernist ideology," rooted in a confidence that centralized planning could bring about scientific and technical progress, limitless expansion of production, and rationalized control of nature.[18] Fishery policy in twentieth-century China was devoted to removing every obstacle to efficient production, thereby enhancing official supervision, ensuring a predictable supply of fish, and obtaining a steady flow of revenues. During the Republican period, state-led developmental plans for Zhoushan's fisheries went hand in hand with the imposition of new frameworks for extraction. This "discourse of development intervention" served as the grounds for extending the power of state agencies and furthering their political aims of control.[19] At the same time, these interventions revealed fissures within the state, as intense conflict arose among agencies competing with one another for revenues.

At best, the institutions formulated by native-place groups in Zhoushan were an incomplete solution to the problems characterizing common-pool resources. Customary regulation limited violent and costly disputes but not increasing pressures on finite marine resources. On the international level, Japan's mechanized fishing fleet entered waters off Zhoushan in the 1920s after exhausting fishery stocks in the East China Sea. Foreign competitors overrode the institutions that Chinese fishers used to regulate common-pool resources and accelerated resource depletion. Internally, population mobility and growing market demand heightened competition and led to a breakdown in arrangements allocating access to fishing grounds, giving rise to prolonged, violent conflicts in the 1920s and 1930s. Local forms of regulation that existed in Zhoushan's fisheries thus proved incapable of coping with large-scale environmental challenges.

Nevertheless, modern assumptions about the human ability to manipulate and control nature prevailing before and after 1949 had even more disastrous environmental consequences. During the Republican period, the Chinese state's efforts to regulate Zhoushan's fisheries and collect taxes via multiple, competing administrative agencies gave rise to interbureaucratic struggles for the control of resources. When successfully put into practice under the PRC government in the 1950s, developmental projects seeking to maximize economic production and fiscal extraction led to the eventual collapse of Zhoushan's most commercially important fish species by the 1970s. Demands on the marine environment derived from forces that were transnational as well as local, bureaucratic as well as socioeconomic. Ultimately they proved detrimental to everyone, especially fish and other marine life.

Tracing Ecological Change Under the Sea: Methodology and Sources

Assessing ecological trends in marine fisheries entails considerable difficulty. The abundance of fish stocks may fluctuate unpredictably due to environmental factors, such as climate change and oceanographic events, which are largely independent of human action. This environmental uncertainty makes it difficult to determine whether changes in the numbers of fish result from naturally occurring trends or from fishing activities.[20] Against the background of these natural fluctuations, the human exploitation of fisheries tends to display a persistent pattern of boom and bust.[21] When a potentially valuable fishery resource is first discovered, people begin to exploit it, and others who see its profitability soon join them. Too many fishers chase a limited amount of fish, causing catch rates and profits to fall off. Eventually, harvesting exceeds the fish stock's biological capacity for replenishment. Returns decline even further as fishers compete for a strained resource base. Once the fishery no longer yields sufficient profits, fishers who have the ability shift their attention elsewhere. Others can only stay behind to fish the depleted stocks. Without reductions in fishing pressure, the resource ultimately declines and collapses.[22]

The warning signs that escalating human demands on fish populations have exceeded their ecological sustainability usually take two telltale forms. First, in response to reduced returns from overexploited fishing grounds, fishers work harder, turn to more intensive fishing technologies, and travel greater distances to maintain or increase their catch rates. Older inshore fishing grounds decline, and fishers move into new, more productive ones farther from shore. From this perspective, the exploration of fishing grounds in more distant waters "is really a measure of exploitation, of a progress towards overfishing."[23] Second, the average size and age of fish decrease, as fishing removes older, larger fish and pressure on stocks does not permit small fish to reach full maturity. Along with declining abundance, the average size of fish steadily decreases.[24]

Unfortunately, reliable statistical data on the productivity of the Zhoushan Archipelago's fisheries does not exist prior to the 1950s. Nevertheless, documentation related to changes in the distribution, abundance, and size of fish species can be found in a variety of historical records. Qing and Republican period gazetteers from the Zhoushan Archipelago and other coastal regions of southeast China contain valuable sections related to fisheries.

Local newspapers and economic documents held at archives and libraries in Mainland China and Taiwan also provide rich information on long-term trends in fishery production and the distribution of fishing grounds. Furthermore, Chinese and Japanese fishery experts carried out detailed investigations of Zhoushan's fisheries during the 1920s and 1930s.

These sources vary widely in precision and clarity. But when read against one another and in light of theoretical literature on fishery science and ecology, they offer rich information on past changes in Zhoushan's marine ecosystem. An equally diverse array of sources demonstrate how people generated and responded to these environmental transformations. In addition to local gazetteers, archival documents, fishery investigations, and oral history collections, temple inscriptions uncovered through fieldwork in the Zhoushan Archipelago offer important sources on local society. These diverse sources make it possible to reconstruct changing relationships between nature and society in Zhoushan's fisheries.

ONE

Migration, Markets, and Marine Life Under the Late Qing

The Zhoushan Archipelago's fisheries experienced a prolonged transformation over the course of the Qing dynasty that coincided with ecological changes throughout China during this period. As China's population boomed from the late seventeenth to the nineteenth century, people migrated from densely inhabited regions to previously unexploited environmental frontiers. As part of this empire-wide movement of people, increasing numbers of coastal households ventured to the islands of the Zhoushan Archipelago. An intricate web of commercial relations and marketing systems made it possible for the burgeoning population to take advantage of the opportunities afforded by Zhoushan's abundant fish stocks. Social and economic trends emerging during the Qing dynasty thus facilitated the expansion of Zhoushan's fishing industry, but over time they also created the potential for violent conflicts over common resources. This chapter begins with a brief sketch of the Zhoushan Archipelago's natural environment and its marine organisms. From there, it turns to the patterns of migration and commercial activity that facilitated the exploitation of the marine ecosystem for human benefit.

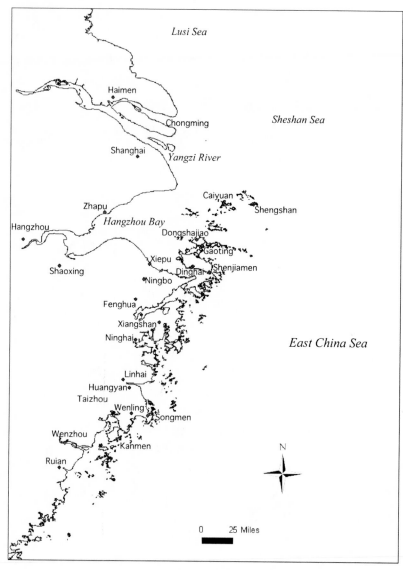

Map 1 The Zhoushan Archipelago and Coastal Zhejiang and Jiangsu.

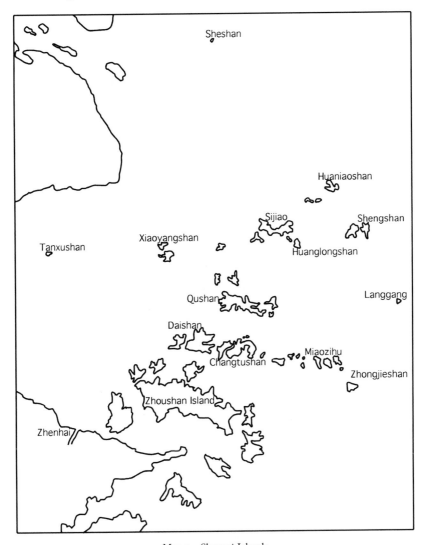

Map 2   Shengsi Islands

The Marine Environment

The environmental characteristics of the Zhoushan Archipelago account for the productivity of its fisheries. The Tiantai and Siming mountain ranges extend from the Zhejiang mainland into the East China Sea near Hangzhou Bay and the mouth of the Yangzi

River. This meeting of land and sea forms a chain of almost a thousand islands, varying in size from tiny islets to almost 200 square miles. In the waters around the archipelago, winds and currents interact seasonally to cause deep, cold, nutrient-rich water to rise to the surface in a process known as upwelling. The islands are geographically situated at the confluence of the high-salinity Taiwan Warm Current (a branch of the Kuroshio Current), the cool Yellow Sea water mass, and the brackish coastal current formed by runoff from the Yangzi and Qiantang rivers. Outflow from the Yangzi River estuary, which empties into the East China Sea north of the islands, provides a generous supply of organic matter.

The front formed by the mixing of these various water systems creates an environment capable of supporting a diverse array of marine life. These waters fertilize plankton that provide abundant nutrition for fish and other forms of marine life, which feed organisms further up the food chain. The character of Zhoushan's coastline, with shallows interrupted by rocky cliffs and a broken seafloor, further contributes to this biodiversity. The combination of these environmental features makes the waters around the Zhoushan Archipelago one of the most productive marine ecosystems in the East China Sea.[1]

During the Qing period, the major commercial species in Zhoushan's fisheries were large yellow croaker (*Pseudosciaena crocea*, Ch: *da huangyu*), small yellow croaker (*Pseudosciaena polyactis*, Ch: *xiao huangyu*), hairtail (*Trichiurus haumela*, Ch: *daiyu*), and cuttlefish (*Sepiella maindroni*, Ch: *moyu* or *wuzei*). Each of these species has long-distance migratory patterns coinciding with seasonal changes in sea temperature. As water temperatures increase during the late spring and early summer, these fish leave their overwintering grounds in the deep offshore waters and swim into shallow coastal waters off the Zhoushan islands to spawn, moving from north to south along the coast of Zhejiang.[2] Seasonal fluctuations in sea temperature correspond to variations in the level of discharge from the Yangzi River estuary and the circulation of the Kuroshio Current. During the spring and summer months, these hydrologic conditions create a warm-temperature, low-salinity environment in waters off the islands that is highly conducive to the production of phytoplankton and zooplankton. The large and small yellow croaker, hairtail, and cuttlefish that migrate to waters off the Zhoushan islands to spawn benefit from high seasonal concentration of these microorganisms, which supports the reproduction and growth of fish species.[3] Cuttlefish enter waters off the islands from late April to early July, hairtail from May to July, small yellow croakers from March to May, and large yellow croakers in May and June. Another, smaller group of large yellow croaker arrives to spawn in autumn. The major fishing seasons in the Zhoushan Archipelago follow these annual migrations.[4]

The Growth of a Migrant Fishery

The exploitation of Zhoushan's fisheries for commercial purposes traces at least to the twelfth century. The increased demand that came with the Song dynasty's relocation to the south led fishermen in Zhoushan to change from gathering fish, shellfish, and aquatic plants for their own consumption to production for the market. As a result, the islands' population expanded, and fishermen started to venture from tidal estuaries to inshore fishing grounds. Most of Zhoushan's commercially important fish species had already come under human exploitation during the Southern Song (1127–1279).[5] Yet fishing boats began to expand into more distant offshore fishing grounds only after patterns of demographic expansion and internal migration placed greater demands on China's environment during the Ming and Qing periods.

China's aggregate population grew substantially during late imperial times (1368–1911). Despite uncertainties about the exact statistics, historians agree that the number of people in China rose from about 100 million during the sixteenth century to over 200 million by the start of the seventeenth century. Decades of natural disaster, civil war, and the Manchu conquest reduced the population by about 40 percent in the mid-1600s. But demographic expansion resumed once the Qing dynasty restored peace and stability. By 1700, population had returned to 150 million. Major growth occurred in the eighteenth century, when population doubled to 300 million. During the 1850s, it reached 380 million. At the beginning of the twentieth century, China's population stood at 500 million and continued to move upward thereafter.[6] In late imperial times, population growth was facilitated by an upsurge in the migration of people into frontier regions. These movements were made possible by the introduction of New World crops such as sweet potatoes and maize that could grow in previously uncultivated areas.[7]

For coastal households, fishing was just one of the livelihood strategies that made this demographic expansion possible. By the 1750s, in parts of northern Zhejiang such as Zhenhai, a coastal county located directly to the southwest of the Zhoushan islands in Ningbo prefecture, residents had opened up highlands and built dikes to turn mudflats into fields.[8] Even with this expansion of cultivated acreage, good land remained in short supply. Families with tiny, marginal landholdings had to engage in additional sideline occupations to ensure their own survival.[9] Consequently, in places like Zhenhai, fishing provided an indispensable means of making a living for families who fished seasonally, as well as for households that engaged in the occupation all year.[10]

Despite its economic importance in coastal regions, China's imperial state looked on fishing with considerable concern, since people who fished were more likely than other groups to join the bands of pirates who plagued coastal provinces throughout the Ming-Qing period.[11] Piracy and fishing had much in common— both required similar skills, and both provided China's coastal residents an important source of income.[12] Hence, as part of its overall policy of coastal pacification, Ming and early Qing rulers placed strict prohibitions on fishing offshore to keep fishermen from collaborating with pirates. On the other hand, officials were also aware that if they eliminated fishing as a source of income, impoverished fishermen would have to turn to piracy to make a

Fig. 1 A traditional fishing boat. From: Zheng, *Chouhai tubian*.

living.[13] For this reason, the Ming and early Qing governments opted to restrict fishing to waters close to shore.

To achieve this goal, official regulations placed strict limits on the size of fishing boats. Before a new ship could be built, its owner had to apply to the local authorities, who would not issue a permit unless the boat's size fell within the proper dimensions.[14] These regulations prevented the construction of large vessels capable of going out to sea for long periods of time. Smaller boats that had to stay within reach of shore were easier to keep under official surveillance and less likely to engage in piracy or other illegal activities such as smuggling. To prevent fishermen from venturing too far from shore, the officials also inspected every boat's grain and water supplies to ensure that they did not exceed the stipulated limits. A mutual responsibility system required each group of ten boats to guarantee one another's conduct after going out to sea.[15] As the historian Ouyang Zongshu suggests, these official restrictions on fishing activity had the unintended consequence of protecting the reproduction of fish stocks.[16]

Maritime defense policies enacted under the Ming and Qing also helped shield the Zhoushan Archipelago and its fisheries from the full effects of human exploitation. During the late fourteenth century, the Ming government promulgated a strict policy of evacuating offshore islands known to be havens for pirates. After 1644, the Qing dynasty followed Ming precedents by prohibiting settlement on the islands.[17] Since the prohibited territories included much of the Zhoushan Archipelago, the islands remained closed to settlement during the early Qing.

Fishery production suffered under these prohibitions, but it was not cut off completely. As demographic expansion in coastal fishing villages filled up fishing grounds close to shore, boats began to defy the government's restrictions.[18] Despite the apparent stringency of the government's fishery policies, local officials charged with enforcing these regulations often ignored them. In addition to the sheer difficulty of restricting the movements of a population that outstripped the Qing bureaucracy's enforcement capabilities, local officials often let fishermen violate the central government's fishery regulations in exchange for various informal fees.[19] Officials had good reason to let fishermen flout these bans, since depriving them of this source of profits would result in poverty and social disorder at the local level. Even when the Qing government carried out its massive coastal evacuation to wipe out opposition from the Ming loyalist Zheng Chenggong regime on Taiwan between 1661 and 1684, fishing boats from coastal areas of Zhejiang defied government prohibitions against offshore fishing. During the 1680s, official observers noted that "poor people" from Zhejiang's Zhenhai, Xiangshan, and Fenghua counties who "used the sea for their livelihoods" gathered in Zhoushan's waters during the spring and summer fishing seasons, setting up temporary residences on the islands to process their catch.[20]

The Qing dynasty first lifted its prohibition on settlement in the Zhoushan Archipelago in 1688, although it did not allow migration to some islands until the eighteenth century. With the removal of official restrictions on settlement, people from mainland regions of Zhejiang wasted no time occupying this newly available land. Settlement of the islands was a wave-like process that had begun by the 1730s and expanded outward to smaller and more distant islands by the mid-1800s. During the Qianlong period (1736–96), migrants began to come to Daishan Island from Zhenhai, Cixi, and Yin counties in the vicinity of Ningbo, as well as from areas in Shaoxing prefecture.[21] In Zhenhai county, as one local gazetteer observed during the 1750s, "Many coastal residents rely on fishing as their profession, and their boats are light and fast. They brave dangers and travel to and from remote islands and areas previously unfrequented by human beings."[22]

During the First Opium War (1839–42), the British occupied the port of Dinghai on Zhoushan Island from 1840 to 1841, but this foreign invasion did little to disrupt migration to the archipelago or the growth of the fishing industry. In fact, by diverting trade to the newly opened treaty port of Shanghai, which became an important market for Zhoushan's fish products, the Western incursion indirectly stimulated the expansion of fishery production. When the Taiping Rebellion (1851–64) devastated Ningbo and its environs, residents of northern Zhejiang fled to safety in the islands. Population movement continued after the rebellion, as people from densely populated Ningbo and Shaoxing as well as "land-hungry peasants" from Taizhou and Wenzhou in southern Zhejiang migrated to the Zhoushan Archipelago.[23] The Taiping upheavals also caused many wealthy families to seek refuge in Shanghai, which increased demand for fish in the growing regional metropolis.[24]

In the latter part of the nineteenth century, fishing households from coastal regions of Zhejiang began to take up permanent residence on the offshore islands. Fishing households from Wenzhou, Taizhou, and Ningbo made their way north to the Shengsi Islands, at the northernmost reaches of the archipelago within the boundaries of Jiangsu province.[25] After 1850, migrants from Fujian and Zhejiang's Ruian and Yin counties settled on the Zhongjieshan Islands at the eastern edge of the archipelago.[26] The bulk of these migrants came to the archipelago from marginal areas on the peripheries of what G. William Skinner identifies as China's Lower Yangzi and Southeast Coast macroregions.[27] Even though these coastal regions had little land suitable for agriculture, they were endowed with an abundance of natural ports that made shipping a readily accessible form of transport.[28]

Settlement of the Zhoushan region resembled the earlier movements of shed people (*pengmin*) from China's land-scarce southeastern provinces, whose migration to hilly districts of the Yangzi valley in the 1720s caused widespread deforestation. A further influx of shed people to the hills of Zhejiang and Jiangsu following the Taiping Rebellion led to chronic erosion of surface soils.[29] Large numbers of fishermen as well as shed people came from the densely populated northern Fujian and southern Zhejiang regions.[30] Opening up highland areas and expanding into offshore fishing grounds were analogous responses to intensified pressure on China's natural resources. In each instance, ecological change brought migration into previously unexploited frontiers. Shed people took to the hills; fishing folk took to the sea. Just as reclaiming uncultivated highland areas wrought ecological devastation through deforestation and soil erosion, increased fishing activity over the course of the Qing heightened demands on marine fishing grounds. Indeed, ecological trends on land and in the sea sometimes interacted with one another. Erosion caused by population movement into highland regions caused many lakes and rivers to silt up. As inland bodies of water disappeared, fishermen turned to lucrative marine fishing grounds.[31]

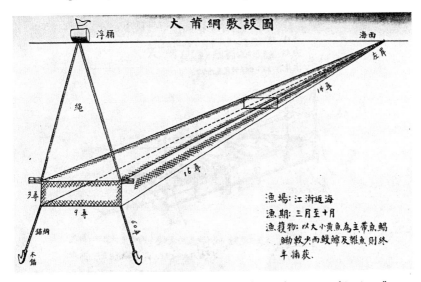

Fig. 2   Two-anchored stow net. From: "Jiang-Zhe qu zhongyao jiushi yuju tu."
February 18, 1948. IMH 20-68b-9-17.

Sojourning fishermen from Zhejiang and Fujian mainland brought bigger and more effective boats and gear from their native regions, which made it possible to target Zhoushan's offshore fishing grounds. Two-anchored stow-net boats (*dabuchuan*) from the ports of Tongzhao and Xifeng on the coast of Xiangshan Bay in southeastern Fenghua county began fishing in Zhoushan's waters in the mid-eighteenth century. This type of fixed gear was particularly well suited for the turbid waters and rapid currents of the seas around Daishan Island.[32] Fishermen from Linhai, Ninghai, and Wenling counties in Taizhou specialized in small pair boats (*xiao duichuan*). This method of fishing used one boat to carry nets and another, called the "feeder boat" (*weichuan*), to direct fishing activities and transport supplies. As early as the fourteenth century, fishermen from Dongqian Lake in Yin county developed a larger version known as the large pair boat (*da duichuan*), which was capable of fishing in more distant waters and staying at sea for longer periods of time. Large pair boats from Dongqian Lake began coming to the Zhoushan Archipelago by the late seventeenth century.[33] Drift-net boats, which were first

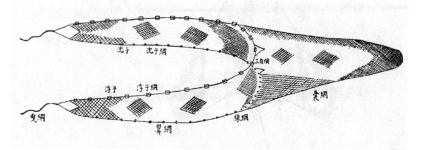

Fig. 3   Pair net. From: "Jiang-Zhe qu zhongyao jiushi yuju tu."
February 18, 1948. IMH 20-68b-9-17.

introduced by fishermen from Zhenhai county, could stay out at sea from ten days to a month at a time.[34] During the Daoguang period (1821–50), drift-net boats fished in waters around Daishan Island, but by the Tongzhi period (1862–74) they had expanded to the Lüsi Sea off the coast of southern Jiangsu.[35] With larger boats and nets, fishermen moved from inland waters to more distant seas in pursuit of productive fishing grounds.[36] In the 1890s, one American observer noted that the "venturesome people" of the Ningbo region "often go a hundred miles, or more, upon the sea, to reach a favorite fishing bank."[37]

In northern Zhejiang, fishermen were not relegated to the low social status of the boat-dwellers (*danmin*) of Hong Kong and the Guangdong region, who lived on their fishing boats as virtual outcastes.[38] In Zhejiang and Fujian, fishing was one component of the diversified economic strategies that coastal households relied on to make a living. In the hilly areas poorly suited for agriculture around Dongqian Lake, households made most of their income by fishing with large pair boats for eight months out of the year. Yet all the lineages in the vicinity of Dongqian Lake that fished also engaged in commerce, agriculture, and other land-based eco-

Fig. 4   Fish merchants from Huian, Fujian. Meadville, PA: Keystone View
Company, 1928. Prints and Photographs Division,
Library of Congress, LC-USZ62-110730.

nomic pursuits.[39] Yin county's Jiangshan village, on the other
hand, possessed abundant paddy fields. Households engaged in
farming for the most part and fished for just two months during
the spring cuttlefish runs.[40] Living standards were relatively com-
fortable in Yin county and Fenghua, and fishing households there
usually owned a house and sometimes a small piece of land.[41] On
the whole, the limited available sources give the impression that
living standards in coastal fishing villages fell only slightly below
those in primarily agricultural regions.[42]

It is entirely accurate to refer to people who harvested fish in
these regions as fishermen, since all of them were male. Women
did not go to sea; rather, they stayed behind to look after the family fields, attend to household labor, and do work that supported

the fishing industry.[43] As was typical throughout northern Zhejiang, fishing households adhered to a gender division in which women maintained the home for sojourning male kinfolk.[44] With male family members absent for several months of the year, female members of fishing households enjoyed more independence than women in farming villages, but they also shouldered heavier labor responsibilities. Two prose poems from the late Qing give a glimpse of gender relations in fishing villages. The first adopts the voice of a woman whose husband has traveled to the Zhoushan Archipelago for the fishing season:

Where, oh where, to express this lovesickness?
On the sea in a small boat, going far away.
Would that my darling's heart be like a swallow,
coming in the spring; going in the fall, without any delay.

As the poem's author noted, fish merchants traveled to Daishan and Qushan islands during the third lunar month and returned home in the ninth month. The second poem makes it clear that even if women did not fish, they played an important part in processing the catch. It loses something in translation, but the sentiment is clear.

Pulling out a swim bladder over five feet long.
I'll use it to describe my inner thoughts and express them to my darling.
All of life is filled with more intimate feelings,
than could be wrapped around this long pole.[45]

Women also made vital contributions to the fishing industry by manufacturing fishing nets as a form of household production. In fishing villages in Taizhou, Zhenhai, and the Zhoushan islands "men, women, and children of both sexes busily engaged in spinning the hemp into thread and twine, making the nets, washing and dyeing them."[46] In Taizhou, old and young women made and repaired nets. Women purchased the raw materials from dealers who dealt in ramie fibers, weaving nets according to their specifications. Net peddlers (*wangfan*), some of whom were sent out by ramie dealers, came to purchase the finished product at dawn on periodic market days.[47]

A Market at Sea

Merchants sold nets and other products in burgeoning markets that centered on the fishing industry. The growth of Zhoushan's fisheries during the late Qing brought a simultaneous expansion in a wide range of other commercial activities. Trade in Shenjiamen, the archipelago's largest fishing port, began to grow soon after the Qing dynasty lifted its ban on maritime trade. Fishermen consumed some of their own catch but sold most. In the early years of the Jiaqing reign (1796–1820) merchants began transporting fish products from Shenjiamen to various ports in the Lower Yangzi region and bringing back rice and soybeans for sale. By the Tongzhi period, a marketing center had taken shape in Shenjiamen, with numerous stalls where merchants carried out trade set up along the streets.[48] Merchants traveled to Zhoushan from coastal Zhejiang and Fujian, opening businesses catering to the demands of the fishing industry. These enterprises exported fish products and imported commodities such as rice, sugar, cloth, soybean cakes, flour, and oil.[49]

In many instances, a single investor raised the capital necessary to purchase or rent a boat, procure supplies, and hire a crew. In other instances, fishermen formed partnerships that pooled capital and divided profits and losses according to shares.[50] Regardless of how they organized production, most fishing boats did not have the cash on hand to cover all expenses.[51] Instead, they obtained investment capital from fish brokers (*yu hang*) operating in fishing ports. As in other sectors of the late imperial Chinese economy, brokers mediated commercial transactions between primary producers and merchants at higher levels of the marketing system.[52] Fish brokers acted as the middlemen who weighed the catch before passing it on to wholesalers.[53] Because they provided capital to producers in the form of loans, as the Daishan Island gazetteer put it, the Zhoushan Archipelago's fish brokers also "possessed the quality of fishery banks."[54] In this way, brokers linked the many highly disaggregated production units that composed Zhoushan's fishing industry with commercial networks that transformed natural resources into marketable commodities.[55]

By and large fish brokers, like the fishermen, were migrants. During the Qing dynasty, most fish merchants in Daishan's port of Dongshajiao were sojourners from Zhenhai who arrived during the summer and returned home during the winter. By the early twentieth century, many had settled on Daishan Island permanently.[56] Other fish brokers from Shanghai, Zhapu, Ningbo, Fenghua, and Taizhou continued to come to Daishan each spring with money and supplies to loan to fishing boats and returned home when the fishing season ended.[57] Sojourning fish brokers from the port of Chongwu in Fujian also migrated to Zhoushan on a seasonal basis to market the catch landed by boats from their native place well into the twentieth century.[58]

The business cycle in Zhoushan's fisheries bore the imprint of the marine environment's seasonal fluctuations. Since fish species migrated to the islands according to seasonal patterns, small-scale fishing enterprises faced periods when capital expenditures vastly outweighed incomes. Fish had to be caught before they could be sold, but the most pressing need for capital came at the beginning of fishing season when funds were least available. During this time of year, fishing boats and fish merchants alike had to find a way to get money they did not yet possess.[59] For all enterprises involved in Zhoushan's fisheries, reliance on credit was the primary means of overcoming this slippage between environmental and business cycles. The commercial exploitation of marine resources would not have been possible without the existence of adequate financial institutions.

What was true of fish was true of other natural resources as well. In an analysis of deforestation in late imperial China, Mark Elvin has touched on the relationship between financial institutions and environmental change. Elvin hypothesizes that with the emergence of financial institutions, the economic returns obtained from cutting down a tree, selling it, and investing the profits at interest began to outweigh the potential increase in value gained by letting the tree grow.[60] These circumstances created what Elvin terms a "cash-in imperative," in which resources that were not utilized "appeared as income foregone."[61] During the Qing dynasty, financial institutions fueled exploitation of Zhoushan's marine resources, albeit in a slightly different manner, giving undercapitalized fishing enterprises a way to carry out production when they did not have funds available.

Fish brokers in Zhoushan extended credit to fishing boats at the beginning of each season. Since boats, nets, foodstuffs, and other essential materials cost money, fishing required substantial outlays of capital. Fishermen typically entered into credit-debt relationships with brokers from their home region. Native-place ties compensated for the fact that most fishermen had little property to offer as security. In all instances, the unpredictability of the marine environment made loans to fishing boats an extremely risky proposition. A bad fishing season or a disaster at sea could make it impossible for fishermen to repay their debts.[62] Financial agreements between fishermen and brokers gave at least some degree of assurance against these risks. Loan arrangements obligated fishing boats to sell their catch only to the broker who had given them credit. In the event that fishing boats could not pay off their debts, brokers retained exclusive rights to purchase their catch during the following season.[63]

Many brokers did not actually loan fishermen money; rather, they supplied rice, nets, and other goods on consignment. Fish brokers traveled to Ningbo, Shanghai, and other urban centers to purchase these supplies, which they then sold on credit to fishermen.[64] Brokers deducted the amount of the debt and the price of provisions from the value of the catch, along with a set commission of 8–10 percent.[65] The convenience that fishermen gained from being able to get supplies when needed offset the markups that brokers often added to the price of goods.[66]

This supply-lending system made it possible for brokers to form a clientele of boats that provided them access to a steady supply of fish.[67] At the same time, longstanding relationships with brokers from their native region enabled fishermen to reduce the difficulty of marketing fish and obtaining credit.[68] In addition to providing access to the capital and supplies, brokers saved fishing boats the time-consuming task of transporting their catch to market. By acting as middlemen in transactions between producers and wholesalers, brokers also minimized the risk that fishers might not find a buyer for their catch.[69] Doing business with fish brokers made it possible for boats to market fish and return to sea as quickly as possible, thereby facilitating exploitation of the marine environment. Division of labor, in other words, led to higher efficiency and greater pressure on fish.

Biographical information on fish brokers in the Zhoushan region supports Susan Mann's contention that brokerage was an open area for entrepreneurship that could be entered and exited freely as the marketed allowed.[70] Hu Baoquan, the largest fish broker in Sijiao Island's port of Qingsha during the late nineteenth century and into the 1930s affords a particularly detailed glimpse into the background of these merchants. Hu was born in Ningbo in 1862, but he and his parents migrated to Qingsha around 1870. During this time, Qingsha was a remote fishing area with little commercial activity. Taking advantage of this situation, the Hu family peddled cakes and tofu in the port to make a living.[71] Fishing season brought increased demand for these foods, which were staples for the crews of small boats that could not accommodate cooking fires.[72] After several years the Hu family prospered, and they opened up a small tofu shop in Qingsha to sell their wares.

In 1884 Hu Baoquan inherited the enterprise from his parents and managed it together with his wife. Hu's wife possessed considerable business acumen and began to let fishing households purchase goods on credit. Around this time, increasing numbers of stow-net fishermen from coastal Zhejiang began to settle in Qingsha, leading to greater demand for supplies. Responding to this opportunity, Hu and his wife decided to change trades. In 1895 they opened up the Yuansen fish brokerage (Yuansen yuhang) in the building that had housed their original shop. To meet the needs of fishermen, Hu purchased supplies such as bamboo and hemp from Ningbo, Zhapu, and other ports. With his wife's encouragement and assistance, Hu permitted cash-strapped fishermen to purchase these goods on credit. By doing so, they ensured access to a supply of fish and attracted numerous customers. The Yuansen fish brokerage's business reputation far exceeded those of other fish brokerages in Qingsha, many of which suffered losses due to insufficient capital and failed to make good on their financial obligations. As Yuansen's business grew, Hu Baoquan went on to open up branches in Ningbo and other fishing ports in the Zhoushan region during the 1930s. Fishermen could exchange credit certificates issued by the Yuansen fish brokerage for cash in all these locations.[73]

Opening a fish brokerage did not require an extensive operation. To enter the trade, an aspiring entrepreneur needed only a rented building, a few flags to identify the boats with which the brokerage did business, an abacus, and several employees.[74] On Daishan Island, fish brokers maintained credit relationships with fish-processing enterprises. These ties provided brokers with the capital they loaned to fishing boats. In exchange, fish brokers promised processors an agreed-on amount of fish, which they dried and salted before transporting to wholesalers in Ningbo and Shanghai. Other brokers simply processed the fish themselves. After obtaining processed fish from Daishan, wholesalers in Ningbo and Shanghai delivered the products to local dealers or to importers in Hangzhou, Shaoxing, Zhapu, Wenzhou, and inland areas up the Yangzi.[75]

In other instances, fishing boats handed their catch over to ice boats (*bingxian chuan*) that transported it from offshore fishing grounds to wholesalers in Shanghai and Ningbo. The ice used by these boats was produced naturally in the Ningbo region and the Zhoushan islands during the winter and stored in icehouses, which farmers opened as another sideline occupation.[76] During the winter, ice producers flooded the fields around ice houses and harvested the ice. A thick layer of straw insulation protected the ice in the ice houses. Relying on this method, wrote Robert Fortune in the 1840s, the Chinese peasant producer "with little expense in building his ice-house, and an economical mode of filling it, manages to secure an abundant supply for preserving his fish during the hot summer months."[77]

Most ice boats had little capital and relied on loans from fish brokers and wholesalers to carry out their business. After a fishing boat came to port, employees from the fish brokerage with which a particular ice boat did business went out with a set of scales to weigh the catch. The boat then loaded its fish onto the ice boat without having to go ashore. Some ice boats met the fishing boats at sea to make their purchases. Rather than paying cash, most ice boats issued credit certificates (*shuipiao*) that indicated the amount and price of the fish sold. Fishing boats redeemed these certificates at brokerages or native banks (*qianzhuang*) after returning to port.[78] Once merchants sold the fish on the market, they transferred funds to the fish brokers' native bank, which remitted funds to a specified native bank or shop, where fishing boats could claim payment for the catch.[79] Fish merchants refined these credit practices during the Republican period by having fishermen and ice boats carry an account book, in which they recorded the amount of fish sold before embossing it with the broker's seal. Fishermen then settled their accounts with the broker at a specific time and location at the end of the fishing season.[80]

Ice boats were vital for getting the catch from fishing grounds to the market in good condition.[81] As G. R. G. Worcester observed, ice boats were:

comparatively modern; for as competition and the greater distances to be covered in search of fish forced the fishing fleets to go farther afield, a means had to be found to keep the fish fresh while it was being transported over the comparatively long distances to the markets of Ningpo and Shanghai. The duty of the ice-boat, therefore, is to follow the fishing fleets, to buy the catch, and to bring it back between layers of ice.[82]

The number of ice boats in the Zhoushan Archipelago multiplied as heightened competition for resources forced fishermen to venture into more distant waters.[83] As fishing grounds shifted away from shore, fish brokers also sent ice boats to transport supplies to fishing boats, making it possible for fishermen to stay at sea for longer periods of time.[84] This method of transporting frozen fish to market functioned as a distribution network that linked producers and retailers, thereby increasing the marketing capacity of Zhoushan's fisheries.[85] As early as 1906, the market for frozen fish in Shanghai "grew ever more prosperous and ice boats coming to port increased with each passing day."[86]

Due to the threat of spoilage, ice boats had to unload the frozen fish they carried immediately after coming to port. Their inability to wait to make a sale exposed ice boats to the vulnerability that came with constant fluctuations in the market price of fish.[87] If ice boats could not get the catch to market in time or could not get a decent price, fish brokers did not see a return either. When this occurred, brokers sometimes went bankrupt, and fishing boats did not receive payment for their catch.[88] In contrast, transport boats filled with salt faced far less uncertainty than ice boats.[89] Salted fish kept longer than frozen fish, and salt boats could wait to sell at a better price. Because salted fish was of inferior quality and fetched a lower price on the market than frozen fish, however, salt boats did not have the potential for large profits that ice boats did. Salt boats usually came to fishing grounds after ice boats left and purchased leftover fish at lower prices.[90]

To minimize the risk and maximize profit, brokers sought to buy the catch from fishing boats at the lowest possible price. Brokers oftentimes did so by manipulating scales when weighing the catch and by purchasing fish at prices below those prevailing on the market.[91] Fishermen sold their catch to brokers "without being informed of the market price, so that brokers are able to monopolize the profits."[92] When fishing boats received loans, brokers deducted a commission double the rate charged on the Ningbo financial market for converting the silver taels used for financial transactions into the silver dollars that served as China's currency. Fish brokers withheld no such premium when calculating payments owed to them by fishing boats. These various forms of squeeze earned fish brokers the disparaging nickname "six-four brokers" (*liu si hang*), since 40 percent of the value of the catch supposedly ended up as their profits, and fishermen only received 60 percent.[93] Fishing boats needed to unload their catch as soon as they could, and wrangling with ice boats and brokers would only result in unwanted delays.[94] Selling at fish brokers' low prices was a better option than running the risk of finding no buyers at all. For this reason, selling through brokers presented the best available avenue for fishermen to market their catch.

Conclusion

Over the course of the Qing dynasty, China's growing population migrated into previously unexploited ecological frontiers in pursuit of new ways of making a living. Whether on land or at sea, these movements of people carried significant environmental consequences. In the Zhoushan Archipelago, patterns of in-migration and growing integration with China's burgeoning commercial economy placed heightened demands on marine resources. From the 1730s through the late 1800s, migration from the densely populated coastal regions of Zhejiang and Fujian forged closer ties between Zhoushan's marine ecosystem and China's commercial economy, giving people opportunities to profit from common-pool fishery resources through their entrepreneurial pursuits. Business activity followed the rhythms of nature, as sojourning fishermen and merchants flocked to the islands during annual spawning runs. Fishing enterprises in Zhoushan were small-scale organizations with limited capital, dependent on brokers who linked them to the larger marketing system and gave them access to credit. Loan-debt relations synchronized economic and environmental cycles, allowing small-scale fishing enterprises with limited finances to catch and purchase fish when they did not have funds on hand.[95] As demographic growth, migration, and commercialization made it easier to convert fish stocks into economic profits, fishing enterprises rushed to take advantage of Zhoushan's marine resources. With no form of regulation, however, the costs of unrestrained competition and violent disputes over fishing grounds would have threatened to diminish this source of income. The following chapter turns to the institutional strategies that Zhoushan's fishing population developed to cope with these interlocking economic and environmental changes.

TWO

Social Organization and Fishery Regulation, 1800–1911

As pressure on Zhoushan's fisheries grew during the late Qing, regional coalitions of migrants from different regions of Zhejiang and Fujian established proprietary claims to fishing territories and tried to limit entry by outsiders. Similar organizational strategies existed among migrant communities in China's urban centers, where native-place groups claimed specific market sectors and tried to regulate trading behavior within them. By dividing the commons into sectors, migrant groups in Zhoushan's fisheries minimized the loss of profits to both unchecked competition and conflict over resources. By ensuring peaceful exploitation, these local forms of regulation prevented costly violence. Local officials granted support and recognition to these arrangements, since during the late nineteenth century they increasingly relied on native-place organizations and their elite leaders to collect taxes and maintain social order in remote maritime fishing centers.

The institutions that regulated the use of Zhoushan's marine resources were inseparable from migrant fishing communities' ritual observances. Deities honored in local temples had the power to grant fishermen security from maritime hazards and risks. By coordinating the use of fishing grounds, regulations connected with these temples likewise safeguarded fishermen's collective welfare in this chaotic and unpredictable world. For local elites and officials, these institutional arrangements minimized social disorder arising from competition for resources. Although regulations effectively minimized conflict, they in no way limited exploitation of natural resources. These arrangements, like fishing communities' religious observances, were concerned with promoting human welfare rather than preserving marine life.

Fishing Groups

Like other groups in Chinese society who left home to labor and do business in unfamiliar regions, the fishermen who migrated to Zhoushan turned to native-place relationships as the main locus of collective affiliation and mutual assistance. As Philip Kuhn has argued, different native-place groups formed a symbiotic relationship that facilitated efficient use of resource space and helped minimize conflict.[1] As fishery production in Zhoushan expanded during the late Qing, growing competition gave rise to frequent disputes over fishing grounds, anchorages, and land for processing the catch and repairing nets. Fishermen from the same native place formed regionally based fishing groups (*yu bang*) to maintain their share of these finite environmental resources.[2]

Sojourning fishing groups in Zhoushan relied on an "ad hoc, quasi-legal" system of fishery regulation based on monopolization of access to fishing space.[3] The regionally based fishing groups staked exclusive claims to particular fishing grounds and anchorages and sought to prevent encroachment by others. In the waters off the port of Dongshajiao on Daishan Island, for example, customary agreements assigned different anchorages to fishing boats from each native-place coalition.[4] The earliest migrants claimed the most attractive fishing spots and passed them on to later generations. Boats using floating gear claimed waters a certain distance away from islands, and fishermen who used fixed nets set up stakes in shallow waters to demarcate their fishing spots. Fishermen were quick to act against violators of these arrangements.[5]

This system divided fishing space according to technologies best suited to particular environmental niches. A Japanese investigation of Zhoushan's fisheries from the early 1920s stated: "There are customs of exclusive use in fishing grounds that are defined according to types of fishing gear and cannot be violated."[6] Regional ties often overlapped technological specialization, since fishermen from different native-place groups specialized in the same types of boats and nets.[7] In Zhoushan's fisheries, "No official restrictions are placed on the size of the mesh or the kind of gear used, but each convoy or fishing fleet has its own particular beat from which, according to local custom, it must not stray."[8] This system of "sea tenure" reduced uncertainty by allocating privileged rights to fishing territories and limiting competition.[9]

Local Religion and Resource Management Institutions

As Robert Weller has argued, premodern Chinese perceptions of the environment did not draw a fundamental distinction between the natural, human, and divine worlds. The same cosmic energy (*qi*) permeated and vitalized the environment, people, and the gods they worshipped.[10] To fishermen in Zhoushan, the most powerful of these deities was the Dragon God of the Sea (Hai Longwang), who ruled the ocean with the "fish soldiers and shrimp generals" (*yu bing xia jiang*) as his subordinates. Whether boats caught fish depended on the Dragon God's commands. According to a folksong from the Zhoushan region, fish ended up in fishermen's nets when the Dragon God made them "descend to the mortal world" (*xia fan*). These unlucky fish could fault only the Dragon God for their misfortune. As one fishing song from the Zhoushan region put it, "You can blame it only on your bitter fate, for when you offend the Dragon God, there is no escape."[11]

As the deity with power over the sea and its inhabitants, the Dragon God held fishermen's fortunes in his hands as well. For fishermen, the marine environment was fraught with risk. If fish did not arrive at the expected place and time, fishermen's nets stayed empty. Accidents caused by the weather and waves were often fatal. Beliefs surrounding the Dragon God expressed fishermen's ambivalent relationship with the marine environment. The deity, like the sea itself, could bring fortune or misfortune. On one hand, because the Dragon God controlled the seas and all types of marine life, fishermen relied on his blessings to make a living. On the other hand, fishermen feared the Dragon God because of his capacity to bring harm. When fishing boats encountered storms or did not bring in a good catch, it was because the Dragon God was angry.[12] The lyrics of another folksong recorded in the early twentieth century expressed this ambivalence.

The sky and the sea are vast and boundless,
the fishermen have offended the Dragon God.
In the fourth and the fifth months they cannot return home,
so anxious that their eyes brim with tears.
Begging the Dragon God to show mercy,
and save this ill-fated young man.[13]

Knowing what could happen if the Dragon God grew unhappy, fishermen carried out ritual observances to please him. When the fishing season began, fishermen made offerings to placate the Dragon God, avoid calamities, and ask for an abundant catch. During the Thanking the Ocean (*xieyang*) festival held at the end of fishing season each summer, fishermen paid their respects to the Dragon God for allowing them to remain safe and bring in a good catch. Just as Chinese farmers prayed to the Dragon God to bring rain during times of drought, fishermen held a ceremonial procession and made offerings to ask for the deity's assistance when the fish did not run. If the Dragon God responded to their requests and the fish arrived, fishermen piled up their catch in front of the deity as an offering of thanks. If the fish still failed to appear, fishermen hung their empty nets in front of the Dragon God's temple to show their dissatisfaction.[14]

Fishing communities had a far less ambiguous relationship with patron deities that preserved their collective welfare. Among these gods was Yang Fu, whose temples existed throughout northern Zhejiang's Ningbo region and the Zhoushan Archipelago.

According to a temple inscription recorded in a late Qing gazetteer from Yin county, Yang Fu was originally a Tang dynasty official named Yang Zhuan, who was the prefect of Mingzhou (the administrative jurisdiction that later became Ningbo prefecture) from 881 to 885. During his tenure, Yang successfully turned back an "invasion" by bandits from southern Zhejiang's Taizhou region and oversaw the construction of fortified walls around the prefectural seat. By the late Ming dynasty, the inscription claimed, local residents had elevated Yang from a gifted official to a deity capable of bringing good fortune, eliminating danger, and providing relief from disaster.[15]

Yet not everyone seems to have understood Yang Fu in this way. Fishermen's perception of the deity diverged sharply from the "official" version of his origins. Oral traditions from the Zhoushan region remember the god not as a gifted official but as a fishing boat skipper surnamed Yang who saved numerous people at sea during his lifetime. After his death, the Jade Emperor (Yu di) made Yang the god in charge of life and death at sea. Fishermen and other seafarers from the Ningbo region saw Yang Fu as their protective deity, regarding him not as "Prefect Yang" (Yang fu) but as "Man of the Sea" (*yang fu*)—two terms with similar pronunciations.[16] Yang Fu's dual identity as fishermen's protective deity and a god of officially respectable social origins resembled that of the goddess Tian Hou (Empress of Heaven), also called Mazu, a better-known deity worshipped by seafarers in Zhoushan and areas all along China's southeast coast.[17] Much as was true of Tian Hou, the meanings that people attached to Yang Fu differed with their social status. Literate elites viewed Yang Fu as a symbol of order and stability. To fishermen, the deity promised protection and security from the hazards and risks of the marine environment.

Fishermen from the Ningbo region of northern Zhejiang brought their patron deity with them when they migrated to the Zhoushan islands during the late Qing. Each year during fishing season, fishermen gathered at temples dedicated to Yang Fu to burn incense and pray for good weather and an abundant catch.[18]

Fig. 5   Opera stage in the Empress of Heaven Temple in the port of
Caiyuan on Sijiao Island. Photo by the author.

As a Qing dynasty poem describing the start of fishing season in
Ningbo put it,

Boats arrive to the sound of beating gongs.
In the early season many large yellow croakers are caught.
First buy a sheep to offer at Yang [Fu]'s temple,
with Yiyang opera performers singing and twirling.

(Yellow croakers caught at the start of the fifth month are called early-
season fish. At the temple of Prefect Yang those who go out to sea
make offerings in great abundance).[19]

Yang Fu's power and efficacy (*ling*) derived from the god's ability
to grant fishing communities welfare and security in a dangerous,
unpredictable environment. Fishermen honored Yang Fu and
asked him to safeguard them and promote good fortune. In return for this divine assistance, fishing communities repaid the
god with offerings and entertained him with opera performances.

Fig. 6   Yang Fu Temple in Dongshajiao, Daishan Island. Photo by the author.

In addition to being the site of collective rituals, local temples provided the public space in which fishing communities enacted regulations governing the use of common-pool marine resources during the late Qing. The clearest example of these institutions is an 1844 inscription headed "Fishing Season Prohibitions" (*Yuxun jinyue*), which still stands in the Yang Fu Temple in the fishing port of Dongshajiao on Daishan Island. These rules applied to migrants from the Ningbo region who gathered at the Yang Fu Temple during fishing season starting in the late eighteenth century.[20] Placing the regulations in Yang Fu's presence implied that they were the deity's prohibitions and imbued them with a sense of impartiality and divine authority.

The Fishing Season Prohibitions coped with problems associated with the technology used by fishermen in Daishan. During the mid-nineteenth century, most of the boats in Daishan fished with stow nets (*zhangwang*). Wooden posts attached these nets to the ocean floor in areas where fish gathered, and a bamboo frame

kept the net open. Ocean currents drove fish into the net, causing it to close and trap the fish inside. Fishermen returned to collect their catch from the nets one to three times a day.[21] Stow
nets were usually quite effective, but they were not without problems. First, strong currents or bad weather might detach nets
from their moorings and wash them away. Second, because the
nets were affixed to the seafloor, boats sometimes collided with
and damaged other fishermen's nets. Finally, someone might take
unattended nets along with the fish inside. The Fishing Season
Prohibitions spelled out a system of rules to avoid the disagreements and conflict that resulted from these problems.

The prohibitions assessed each fishing boat 500 copper cash to
meet expenses for opera performances and banquets held to en-
tertain Yang Fu, but which fishermen enjoyed as well. Any fishing
boat that damaged others' nets had to pay 100 copper cash per
net as "apology money" (*xieqian*) to the boat that incurred the loss.
If the damaged nets were recovered, the boat that had suffered
the loss was to pay 50 copper cash out of the apology money to
the finder. The regulations also required fishing boats to set up
posts at the spot where nets were lost or found as verification.
Fishermen who found nets had to report their discovery to the
rest of the community immediately after coming ashore and
could not keep it a secret. If fishermen were discovered trying to
conceal nets they found, the nets would be reclaimed and the
apology money would be confiscated (*chong gong*). The offender
would also have to pay for one opera performance as a fine. The
individual who reported the guilty party received 20,000 copper
coins out of an additional fine of 60,000 copper coins. During
foggy weather, boats could gather their nets but could not put
down new anchors. This restriction was intended to avert conflicts that could result from tangled gear or collisions between
boats. Boats that violated prohibitions against fishing in poor
weather would also be fined one opera performance. Any fishing
household that failed to attend meetings held to deliberate on
punishments for breaking the rules would suffer a fine as well.

Finally, if violators resorted to force or refused to abide by punishments spelled out in the regulations, then officials would be called on to deal with them.

The Fishing Season Prohibitions address technological externalities that result when fishers physically interfere with one another during their harvesting activities.[22] By spelling out a system of sanctions and compensation, the regulations sought to minimize problems that could arise when fishermen damaged others' gear. Without such rules, unrestrained competition would have resulted in the widespread destruction of fishing gear and physical conflict. The Fishing Season Prohibitions also indicate that during the mid-nineteenth century fishing groups had the authority to impose sanctions to enforce these regulations and looked to official intervention only as a last resort.

Local folklore records vivid descriptions of these punishments. If fishermen caught someone who had stolen or damaged others' nets, they tied the offender to the posts that secured nets to the ocean floor and waited for him to drown. If the perpetrator committed the offense against a fisherman from his own kinship group, lineage elders would hold an assembly to decide on the proper penalty. The guilty party received a beating for a minor offense or the death penalty for a major one. If the perpetrator somehow got away, fishermen made a straw effigy, which they strung up and flogged in his place before throwing it into the latrine. This symbolic penalty sent an unambiguous message. When conflicts occurred among fishermen, a respected local elder would conduct mediation. The guilty party would have to pay for an opera performance as a fine, thereby distinguishing right from wrong and setting a clear example for others.[23]

In general, people are able to deal with common-pool resource dilemmas most effectively if they are part of a community that gives them opportunities to discuss their problems and to propose solutions.[24] In Zhoushan's fisheries during the late nineteenth century, the temple community gave native-place coalitions a setting for interaction in which they could work out rules coordinating the use of common resources. Religious rituals held at local temples also defined the native-place community, situating it as a collective entity distinct from other groups. As in the irrigation communities in north China studied by Prasenjit Duara, temples in Zhoushan functioned as the apparatus of coordination and control necessary for the coordination of environmental resources.[25]

Placing the inscription bearing the Fishing Season Prohibitions in the temple that represented the nexus of social activity and public interaction in the fishing port ensured their transmission and dissemination. Most of the fishermen to whom the Fishing Season Prohibitions applied could not have understood the regulations in their written form. Literate members of the local elite would have had to read the prohibitions aloud to illiterate fishermen when they gathered in the temple to make offerings during fishing season. As a result, fishermen would have been made continually aware of these rules and the consequences of violating them. Respected and capable local elites also assumed leadership roles in the associations that managed temple affairs and oversaw festivals and other ritual observances. The authority and prestige that these elites gained by serving the deity and the larger native-place community enhanced their ability to mediate disputes over fishing grounds. Like the deity honored in the temple, the regulations held different meanings for different social groups. To fishermen, the prohibitions granted security against the hazards and risks that accompanied their trade. To elites, the rules maintained order and stability within the native-place community by minimizing violent conflict over common natural resources.

Fishing Lodges

The elites who had the power, prestige, and influence to manage local temple associations also acted as the leaders of fishing lodges (*yuye gongsuo*) formed by fishermen and merchants from each native-place group. In Zhoushan, as in other areas of China, the term *gongsuo* (literally "public office") was used interchangeably with the term *huiguan* to refer to native-place and same-trade associations.[26] Like other native-place organizations formed by merchants and artisans throughout China during the Qing period, fishing lodges were made up of sojourners from specific geographical areas. The first recorded fishing lodge in the Zhoushan region came into existence in 1724 when stow-net fishers from Dinghai and Zhenhai established the Nanpu Lodge in Ningbo.[27] The temporal pattern of fishing lodges' formation paralleled patterns of in-migration in Zhoushan's fisheries, as well as the competition for resources that they entailed. Migrant fishermen and merchants founded only a handful of fishing lodges prior to 1850, but their numbers expanded greatly during the late nineteenth and early twentieth centuries.[28]

Considerable overlap existed between local temples and regional fishing lodges. Fishermen from each native-place group gathered to worship at certain temples, which frequently doubled as headquarters for lodge organizations.[29] Overlapping local temples and fishing lodges were not just spaces for interaction. They were also sites of demarcation that divided one native-place community from others. In this respect, fishing lodges resembled native-place associations throughout China during the Qing; all of them functioned as spaces for particularistic regional groups to conduct ritual observances in honor of their respective patron deities and other gods. Given these parallels, it comes as no surprise that the title used to refer to fishing lodge headmen (*zhushou*) also referred to local notables who organized and administered temple rituals, festivals, and other collective activities.[30]

Like the temples with which they were connected, fishing lodge organizations provided vital ritual services for migrants from their native-place collective. The Taihe Lodge formed by fishermen from Xiangshan had a charitable cemetery where deceased members "rested" before returning to their native place for burial. The lodge's members gathered to make offerings to their departed compatriots during the Thanking the Ocean festival at the end of fishing season. For individuals lost at sea, fishermen performed a ritual at high tide asking the Dragon God of the Sea to let the soul return to shore and enter a paper effigy of the deceased, which was then buried as a substitute.[31]

If the burial rituals were not performed, the drowned fisherman's soul would wander the sea as a water ghost (*shuigui*). According to popular beliefs, these spirits threatened the safety of other fishermen. "Having spent some time in their wet abode in the bondage of the watergods, they may be redeemed from this servitude by substitution, and therefore they lie in ambush for victims to draw into the water and make them take their place."[32] An inscription dated 1887 and found in Daishan Island's Yang Fu Temple records that fishing households in the Anlai Lodge made payments to the families of deceased crew members to cover rituals to call back the soul and for burial. According to another inscription from the fishing port of Dongmen, if fishing households did not provide funds for the calling back the soul ritual, "local bullies" would seize the opportunity to urge the deceased fisherman's widow to "fabricate stories and act crazy," thereby giving them an opportunity to file lawsuits.[33] By mediating between the native-place community and marginal, antisocial elements (both human and nonhuman), the calling back the soul ritual reinforced collective welfare and social cohesion against threatening outside forces. Collective religious observances, as Susan Naquin points out, forged a ritual sense of public belonging that reinforced shared native-place sentiments.[34]

Merchants who ran fish-processing facilities on Daishan Island formed two distinct groups defined according to native-place. The "guest group" (*ke bang*) consisted of merchants from Zhenhai county who rented land in Daishan to process the catch during fishing season. Most of the merchants in this group came during the summer and returned home in the winter, but some resided in Daishan year-round. The "local group" (*tu bang*) consisted entirely of merchants who lived permanently on the island. During the eighteenth century, sojourning merchants from Zhenhai started the first fish-processing enterprises on the island, and this group was also known as the "old fish merchants." The group that had settled permanently was known as the "new fish merchants."[35] In 1796 fishing households from Beixiang village in Zhenhai county established the Fish Merchants Lodge (Yushang gongsuo). Originally, this lodge's membership was made up of fishing households (*yuhu*) who caught and processed fish. As a clearer division of labor took shape in the late Qing, the lodge came to be composed entirely of merchants who specialized in processing fish. The number of locally run processing enterprises later began to expand, and in 1887 fish merchants who settled on Daishan established a New Fish Merchants Lodge (Xin yushang gongsuo). When the new lodge was formed, the Zhenhai merchants' lodge was referred to as the Old Fish Merchants Lodge (Lao yushang gongsuo).[36]

Merchants and gentry from each regional group filled fishing lodges' leadership positions.[37] Lodge organizations possessed several internal subdivisions (*zhu*), the exact number of which varied according to their size. Each of these subunits selected one or two headmen (*zhushou*) who acted as their representatives and contacted elites from their native region to handle the lodge's affairs.[38] During the 1910s, according to the Daishan gazetteer, "Each fishing fleet has established a lodge in Dongshajiao. They draw lots to appoint chiefs and extend invitations to directors (*dongshi*)."[39] The 1915 regulations of the Yongan Lodge, which was formed by large pair boats from Yin county, stated that fishermen selected a number of headmen to act as the organization's governing body. These individuals then assembled to elect a director and appoint a manager (*sishi jingli*).[40] Elite leaders gave fishing lodges the funding needed to underwrite their activities, as well as vital connections with government officials.[41]

Fishing Lodges and the Maintenance of Local Order

When necessary, lodges mediated arguments that arose during commercial transactions between fishers and fish merchants, arbitrating disagreements over prices, weights and measures, and commissions.[42] If disputes broke out among fishermen, "the lodge leaders (*gongsuo xiansheng*) settle them justly and no one is dissatisfied."[43] Like same-trade associations in virtually every sector of the Chinese economy, fishing lodges functioned to minimize the negative consequences of unrestrained competition between members of their trade.

As early as the Kangxi period (1662–1722), the Dinghai county magistrate recommended using fishermen's native-place groups as the basis for the local government's surveillance and mutual responsibility system. Naval patrols accompanied each fleet to and from fishing grounds, making sure they came back to port every night and returned to their native region at the end of fishing season. Local officials punished the whole fishing group if any boats violated the regulations.[44] During the late nineteenth century, as the Qing state's hold over much of Chinese society declined, local elites assumed greater responsibility for local security and tax collection.[45] The role of regional fishing lodges in Zhoushan during the late Qing reflected this larger trend.

With the outbreak of the Sino-French War in 1884, Zhejiang's provincial governor Zuo Zongtang appointed the Yongan Lodge's director Hua Ziqing general director of a new Zhejiang Fishing Militia (*Zhejiang yutuan*). The founding of this defense force coincided with a general upsurge in formation of militias by local gentry in northern Zhejiang after the outbreak of war with France.[46] The Fishing Militia was made responsible for inspecting fishing boats, compiling mutual-security (*baojia*) registers, issuing permits, and collecting registration fees. The Fishing Militia's regulations stated that any surplus funds could be used as charitable aid for fishermen or to hire mercenary braves for mutual defense. Contrary to official intentions, however, fishing lodges and their gentry managers (*shendong*) manipulated this position to gain the upper hand over their competitors. The Fishing Militia's director funded defensive convoys for boats from his native-place coalition but ignored acts of piracy committed against others. Not surprisingly, these actions met with a wave of protests from other regional fishing groups, and the militia disbanded nine months after its inception.

Following the Sino-Japanese War in 1895, Zhejiang's provincial officials revived the fishing militia system in response to increased piracy along the Ningbo-Taizhou coast, appointing the

Fengnan Lodge's head Liu Xiaosi as director. However, the dispersed character of Zhejiang's fishing population, much of which lived on scattered offshore islands, made it difficult for the Fishing Militia to carry out its surveillance and control functions. Hence, in the early twentieth century the Fishing Militia Bureau (Yutuan ju) turned into an ad hoc office for collecting license fees from fishing boats on local governments' behalf.[47] These fees were among the many commercial taxes through which the Qing tried to broaden and increase revenues to restore imperial power following the domestic rebellions of the nineteenth century. In Zhoushan, as in many other regions of China, collection of these levies was frequently farmed to local elites.

Previously, fees collected from fishing boats were one of the numerous "miscellaneous taxes" (*za shui*) outside the regular fiscal accounts of local governments. Officials and clerks "reported fishing taxes as they pleased." These informal levies often provoked resistance from fishermen's native-place groups, which relied on "capable and experienced gentry" to petition officials for elimination of permit fees. Pressure from their elite leaders helped many fishing groups obtain reductions in these levies.[48] The Fishing Militia Bureau aimed to consolidate official control over revenues from fishing boats by reducing registration fees and eliminating all other informal levies. The Fishing Militia's 1895 regulations granted leaders of fishing lodges responsibility for collecting these taxes.[49] Lodge organizations went on collecting such fees even after the Fishing Militia Bureau dissolved and authority for the registration of fishing boats passed to Zhejiang's provincial coast guard (Waihai shuijing ting) in 1917.[50]

The Ecology of Piracy and Local Security

With the "thousands" of sojourners who came to Daishan Island during fishing season, "the good and bad intermingled." Pirates often ran amok in Dongshajiao's marketplace, robbing people and looting goods. Particularly large numbers of pirates accompanied the fishing fleet from Taizhou.[51] This region of southern Zhejiang's mountainous interior and rugged, island-strewn coastline made it ideal terrain for brigands who, "in conjunction with the island fishermen," attacked trading junks that frequented the ports.[52] Many of the "so-called pirates" from Taizhou were "oftentimes nothing more than fishermen, who have been brought to the verge of destitution by the failures of successive seasons of their legitimate means of livelihood."[53] Many offshore islands near Taizhou became "pirate villages," in which farmers and fishers took part in periodic acts of piracy.[54]

The threat of piracy aroused great fear among the residents of Daishan Island when fishing season came each spring. Many households fled from town whenever word spread that pirates had arrived. Because Daishan was located far away from the Dinghai county seat, however, local officials paid little attention to such lawlessness. The situation changed in 1893, when pirates in Daishan became even more numerous and started to plunder local villages. Incensed local residents resolved to search out pirates and prevent any further incidents. While investigating one residence, a group of locals came across a Taizhou man whose clothing and belongings seemed out of the ordinary. Suspecting that he was a pirate, the locals executed him on the spot. Afterward, they killed three other Taizhou natives and destroyed two buildings. Taizhou people soon enacted their revenge by capturing 300 local households and burning down five buildings in the village of Gongmen.

Following this incident, Daishan natives filed a report with local officials, and the injured Taizhou people also demanded reparations. The costly string of lawsuits that ensued implicated many individuals, including numerous local elites. The dispute went on for over a year before coming to an end in 1894.[55] These events came to the attention of Zhejiang's provincial officials, who stationed local officials in Daishan to inspect strategically important areas during fishing season. Military units were stationed in Daishan, and navy steamships plied the waters around the island to arrest pirates.[56] The provincial authorities also set up a mutual responsibility system for fishermen from Taizhou, in which the native-place group's leaders drew lots to appoint an inspector who punished any illegal activities.[57]

Piracy became especially serious whenever environmental fluctuations led to unusually poor catches. A poem describing the 1906 yellow croaker season in Daishan captured the social and economic consequences of these seemingly random environmental factors in vivid detail:

Under our dynasty, the sea is fortunately peaceful.
The waters are tranquil, and the boats are at ease.
For 100 or so *li* around the Island of Penglai [Daishan],
a year's living is made during fishing season.
The fishing season begins at Mangzhong [in early May],
and a thousand masts, like threads, jam the sea-lanes.
When the first season concludes, the second season commences.
When the yellow croakers come, it is like a surging tide.
Three days after the last quarter of the moon, the second season begins.
At that point, skilled men use poles
and listen for the news [that the fish have come].
The silver waves bring croaking sounds as loud as thunder.
The croaking, like rolling thunder, alerts the fishermen.
Each and every one is exuberant.
Taking up their nets, they can easily make ten million pieces of gold.
Ivory meat and silver scales, as cheap as dirt.
When I came here, why was it not so?
The vast expanses of waters have few fish swimming in them.
Who says the seafood piles up like mountains?
Are the stories not true?
One can only look on and sigh.
Over the past few years, it may be recalled,
there were slight differences between abundance and scarcity,
but nothing like this year's great decline.
Casting nets and coming up with nothing, the sea is full of sadness.
Joy is joy; sorrow is sorrow.
The senior official patrols day and night,
as if there were a great enemy before him.
All of sudden, the great shock awakens you from a dream,
In the kingdom of the sea, they cautiously guard against a great battle.
But with the navy's imposing military prowess,

the war vessels have sent them all away.
Alas!
Ice boats and two-anchored stow-net boats give it a go, but easily
  lose a million.
Only small boats have no way of making a living
and have to engage in illegal activities, which is truly pitiable.
Who has actually made this happen? Only heaven is to blame.[58]

Zhoushan's marine ecosystem, and hence catches, fluctuated considerably from year to year. The available sources do not offer a definitive explanation for the sudden decline in catches in 1906.[59] Whatever their cause, the poem suggests that unexpected fluctuations in catches led some fishermen to turn to piracy and increased the possibility of conflict. Under these circumstances, local officials called on the leaders of fishermen's native-place organizations to maintain order in local society.

Migration, Militarization, and Resource Competition

During the latter part of the nineteenth-century, China grew increasingly militarized as gentry-led militias were formed to combat the Taiping and other domestic rebellions.[60] In parts of the Zhoushan Archipelago, militarization and the formation of mutual-defense forces by regional fishing lodges closely paralleled ecological transformations. As the value of common natural resources, and the possibility of losing them to competitors, grew, native-place groups devoted more energy to protecting their proprietary claims to fishing grounds. As on many other islands in the archipelago, Huanglong Island's fisheries expanded in the 1850s with its settlement by fishermen from Ningbo and Zhenhai.[61] These groups fished with stow nets in shallow waters around the island where stakes could be set up to secure their gear. Stow-net fishermen from the Ningbo region faced competition from boats that came to fish on a seasonal basis from Fujian and southern Zhejiang. Disputes frequently took place between Huanglong's stow-net fishermen and boats from Fujian over accusations of damaging gear and taking fish from unattended nets. These altercations occurred primarily because of interference between the stow-net fishing group's fixed gear and the floating gear used by fishers from Fujian. To prevent losing their catch to Fujianese, in the early 1900s stow-net fishermen and their elite patrons obtained official permission to station patrol vessels in fishing grounds.[62]

After a violent feud between the Huanglong and Fujian groups in the summer of 1905, the Fujianese demanded redress from the Huanglong fishers' Beiyang Lodge. The lodge's handling of the situation only enraged the fishermen from Fujian further, leading them to attack its director. The owner of Huanglong's Shishun-xing fish brokerage (Shishunxing yuhang), a migrant from Ningbo who was the patriarch of Huanglong's wealthiest family, eventually had to pay both sides' medical expenses to bring the conflict to an end.[63] By mediating a settlement, the merchant avoided losing income as a result of the dispute. After this incident, the director of the stow-net fishing group's lodge petitioned government officials in Ningbo for permission to set up a local militia in Huanglong. Additional feuds with fishermen from Fujian and Fenghua motivated village headmen (*zhishou*) in Huanglong to form another self-defense organization called the Gather Knives Society (Longdaohui), which later combined with the local militia to keep order on the island.[64]

The lodge's patrol boat intervened in a dispute between stow-net fishermen and hook-and-line boats from Fujian in 1907, leading to armed fighting in which the Fujianese suffered heavy casualties. The Fujianese boats returned to Shenjiamen and reported the incident to their Bamin Lodge. The organization convened a meeting of representatives from 3,000 boats and resolved to collect funds to file lawsuits against their opponents. The Huanglong group took the same course of action, resulting in a string of lawsuits that ended only after officials intervened.[65]

Following this feud, the lodge outfitted its patrol boat with Western guns and bulletproof steel plates. Every stow-net fishing enterprise in Huanglong paid a fee of three silver dollars per net to cover the expense of hiring a vessel from local government authorities in Ningbo to patrol the island's fishing grounds. This arrangement lasted until around 1921, when conflicts over cutting gear underwent a gradual decline. Instead of investing funds to maintain a patrol vessel, starting from the early 1920s fishing boats formed a village compact (*xianggui*) that required boats found guilty of damaging nets to pay compensation.[66]

Elite Networks and Dispute Mediation: Fenghua Fishermen and the 1911 Revolution

When contests for the control of Zhoushan's fisheries escalated into conflict between fishermen from competing native-place coalitions, fishing lodges' elite leaders mobilized their social networks to conduct mediation and limit economic losses due to these disputes. In the case of fishing groups from Zhejiang's Fenghua county, local conflicts over resources drew in some of the most prominent leaders of China's 1911 Revolution. In this instance, elite groups successfully mobilized fishermen's native-place networks to further their own political agendas.

In early 1911, a violent feud broke out between fishermen from Fenghua's Xifeng group and Ninghai's Zhangshu group over fishing grounds in the vicinity of Qushan Island. The Fenghua fishing group's well-armed self-defense forces killed 25 Ninghai fishermen and left another 36 injured, without incurring any casualties itself. Hostilities persisted after sojourning fishermen left Zhoushan and returned to their home counties at the end of fishing season. To exact revenge against their Fenghua rivals, fishermen from Ninghai set up a lookout post on the border between Ninghai and Fenghua counties. If any people from Fenghua crossed this boundary, Ninghai residents took them captive regardless of age or sex. All in all, 57 individuals from Fenghua were captured in just a few months. Feuding between the Ninghai and Fenghua groups dragged on, and fishermen eventually stopped going out to sea altogether. This abrupt halt in fishing activities caused anxiety not only for local fishing families but also for local governments in Dinghai, Fenghua, and Ninghai counties, which stood to lose an important source of tax revenues.[67]

The fishing dispute soon came to the attention of Ying Meng-qing, a member of the Revolutionary Alliance (Tongmenghui), who was engaged in anti-Qing revolutionary activities in Shanghai.[68] Once Ying received news of the conflict from friends and relatives in his home county, he set about finding a way to settle the feud.[69]

A combination of interests moved Ying Mengqing to seek a resolution to the conflict. On one hand, the hostilities had an adverse effect on the livelihoods of the two counties' residents. On the other, Ying Mengqing had strong economic motives of his own for wanting to settle the dispute. Ying's family lived in Chunhu, a market town located near the port of Xifeng in Fenghua county. The Ying family owned Chunhu's Dechang General Store, which sold ramie fibers for making fishing nets, tong oil used as a protective coating for nets, alcohol, and other goods. Most of the Dechang General Store's business came from fishermen and fish merchants from Xifeng, and the shop's financial fortunes rose and fell with fluctuations in the fishing industry. Hence, Ying Mengqing wanted a prompt resumption of fishing activities in Fenghua to keep his family's store in business.[70]

Ying brought the fishing dispute between the Fenghua and Ninghai groups to the attention of the Revolutionary Alliance's leader Chen Qimei, who wired a returned student from Japan named Chen Yunsheng and Restoration Society (Guangfu hui) member Sun Guansheng, both of whom carried considerable influence in Ninghai. Chen Qimei arranged a meeting with Chen Yunsheng and Sun Guansheng in Ningbo, telling them he had an extremely important matter to discuss. Ying Mengqing and Chen Qimei set off for Ningbo, where they met with Chen Yunsheng, Sun Guansheng, and another Restoration Society member named Shen Changxin. They decided that Chen Yunsheng and Sun Guansheng would first travel to Ninghai and negotiate with the families of fishermen who had been killed and injured in the feud. The others would go to Fenghua to meet with the Xifeng fishermen.[71]

Upon arriving in Fenghua, Ying Mengqing, Chen Qimei, and Shen Changxin assembled local fish brokers and the leaders of the

Shen lineage, which held a dominant interest in Xifeng's fisheries throughout the early twentieth century, at the Shen lineage's ancestral temple.[72] According to Ying's later recollection of the meeting, "They knew me as the young owner (*shao laoban*) of the Dechang General Store who had come back from studying in Japan. They had heard that in order to mediate this case of homicide I had invited some important personages from elsewhere to come along as well. For this reason, they all happily came to the meeting."[73] Chen Qimei gave a speech to the approximately thousand local residents in attendance. Turning to explicitly nationalist rhetoric, Chen explained that Fenghua people and Ninghai people were all Chinese compatriots who needed to unite rather than slaughter one another. Chen urged fishermen in Fenghua to settle their dispute with the Ninghai group and take to the sea at once so they would not miss the fishing season. Chen Qimei also instructed fish merchants from Xifeng to gather funds to compensate the families of the deceased and injured Ninghai fishermen. Since the feud had cut off these merchants' business, they too wanted fishing to resume as soon as possible. The fish merchants approved of Chen Qimei's proposal and appointed representatives to go to Ningbo and negotiate with the Ninghai group.[74]

With matters in Fenghua taken care of, Ying Mengqing went with the other Revolutionary Alliance and Restoration Society members to Zhangshu village in Ninghai, where they met with Chen Yunsheng and Sun Guansheng. The Fenghua elites paid their respects to the families of the deceased fishermen, who agreed to put aside the dispute and resume fishing. The elites also secured the release of the Fenghua residents who had been taken captive and planned for a meeting in Ningbo between the families of the Ninghai fishermen and representatives of the Fenghua group. After making these arrangements, the Revolutionary Alliance and Restoration Society members obtained approval for this settlement from the Fenghua and Ninghai county governments.[75] Local elites who profited from the fishing industry and local governments that taxed it shared an interest in ending the conflict and securing a prompt resumption of fishing activities.

The leaders of the Ninghai and Fenghua groups gathered in Ningbo soon thereafter. Following protracted negotiations, the Fenghua group agreed to pay 500 *yuan* in compensation to the families of each of the deceased Ninghai fishermen and 100 *yuan* to all the injured. Fish merchants from Fenghua bore the financial burden of these payments, which amounted to a total of 15,100 *yuan*. The families of the deceased and injured Ninghai fishermen sent reports to Ninghai, Fenghua, and Dinghai county officials, informing them the feud had come to an amicable resolution and that they would resume fishing at once. Shen Changxin passed the details of the mediation proceedings on to the county governments, all of which voiced their satisfaction. With the conflict over, fishing boats from Ninghai and Fenghua headed back to the Zhoushan Archipelago for the summer fishing season. According to Ying Mengqing, he and Chen Qimei received numerous letters of thanks from fishing households and merchants for bringing the dispute to an end.[76]

Following the Wuchang Uprising in the fall of 1911, Revolutionary Alliance members used connections forged during their mediation of this feud to organize a "dare-to-die squad" composed of fishermen from Fenghua to take part in the anti-Qing insurgency. Elites relied on the same networks that facilitated resolution of fishing disputes to gain the loyalty of local recruits and put together this military force. Ying Mengqing and the other elites who had brokered a settlement to the Fenghua-Ninghai conflict undertook recruiting efforts. After gathering funds in Shanghai, Ying traveled to Ningbo, where he met with Shen Changxin and the Revolutionary Alliance member Chen Xiasheng. Under the false pretense of recruiting railway laborers, they offered 155 fishermen from Xifeng monthly wages of 16 *yuan*, food, and travel expenses, as well as an upfront payment of 10 *yuan* to the families of the recruits. Fishermen who signed up told their families they were going to work as hired laborers to allay their concerns. All in all, 112 fishermen reported for duty in Ningbo five days later.

Ying Mengqing and other anti-Qing revolutionaries brought the fishermen to Shanghai by steamship and transferred them to a train bound for Hangzhou. After their trip, the fishermen found accommodations in the Fenghua Lodge (Fenghua huiguan) and other hotels in the city. Once the military leaders Chiang Kai-shek (Jiang Jieshi), Zhang Boqi, and Wang Jigao had carried out an inspection of the recruits, they were placed under the command of Shen Changxin and Chen Xiasheng.[77] To avoid being discovered by local authorities in Fenghua, Ying Mengqing took up residence at the New Learning Company (Xinxue huishe), a progressive publishing house in Shanghai.[78] Jiang Beiming, a Fenghua compatriot who was one of the New Learning Company's founders, also happened to be the director of Fenghua fishermen's Xifeng Lodge.[79]

Despite the precautions that Revolutionary Alliance members took to elude the notice of the Qing authorities, their activities in Fenghua aroused the suspicion of local officials. Ying Mengqing's elder brother came to Shanghai and informed him that the Fenghua county government suspected he had ulterior motives for recruiting fishermen and had come to the Ying family's home to interrogate him. When government personnel showed up at the New Learning Company in Shanghai to question Ying Mengqing several days later, he claimed that he hired the fishermen to work as laborers on the Shanghai–Hangzhou Railway. As Ying explained, fishermen regularly engaged in temporary employment during the off-season, which made them more than willing to take this job. "It is not fishing season now, and taking advantage of the slow season to do temporary work and earn some wages is also advantageous to the fishermen."[80]

Despite this cover, Ying's father soon came to Shanghai and reproached his son for the troubles that he had created for the family. Fishing households in Fenghua had heard that the fishermen had not gone off to work as railway laborers but to serve as revolutionary soldiers. The Ying family now had to deal with the fishing households' demands that the recruits come back home.

If any of the fishermen died or were injured in battle, their relatives would demand recompense from their elite patrons, just as they did when fishermen perished in accidents. Of even more immediate concern to Ying's father, if the Qing officials discovered the truth of the matter, they would punish him as head of the household. Ying Mengqing convinced his father that the fishermen had really gone to work on the railroad and asked him to assure their families that nothing bad would happen.[81]

Although the anti-Qing revolution met with little resistance in Hangzhou, the Fenghua fishermen's experience defending their turf in Zhoushan's fisheries made them capable soldiers. The fishermen's dare-to-die squad led assaults on the Zhejiang provincial governor's headquarters and the city's armory without incurring any casualties. Ying Mengqing claimed that, after Chiang Kai-shek witnessed the fishermen's performance in battle, he wanted to enlist his follow Fenghua natives in his personal military forces. Chiang commanded his subordinate Wang Wenxiang to take the fishermen to Shanghai, where they would serve in the First Regiment of the Hu Army under Chiang's command. News of Chiang's orders created panic when it reached fishing families in Fenghua. The fishermen's relatives constantly pestered the Ying family to let the soldiers return home. Ying Mengqing's elder brother again went to Shanghai and persuaded him to let the fishermen go back to Fenghua. Ying Mengqing successfully convinced Chiang Kai-shek to demobilize the fishermen. Since most of the fishermen disliked the discipline and lack of freedom that came with military life, they wasted no time in leaving the army.[82]

The participation of Fenghua fishermen in the 1911 Revolution thus stands as an example of elite groups drawing on fishermen's native-place networks for their own political purposes. Using connections forged through the mediation of fishing disputes, reform-minded elites took advantage of local defense forces formed to protect claims to limited resources and temporarily mobilized fishermen from their native region for military action.

Conclusion

As human pressures on Zhoushan's fisheries grew during the Qing, native-place coalitions devised social institutions to coordinate the use of common-pool marine resources and to minimize the costs of violent conflict. Native-place groups limited entry to fishing territories to which they claimed *de facto* proprietary rights and mediated disputes within them. Interpenetrating religious practices and local forms of regulation granted fishermen security from the risks that characterized their trade. Patron deities protected fishermen from the hazards and unpredictability of the marine environment, and rules governing the use of fishing grounds averted chaos and uncertainty due to internecine competition. Local temples functioned as a space for public interaction, as well as a site of demarcation that differentiated particular native-place communities from others. By reproducing collective solidarity, promoting shared values, and legitimizing the power and authority of the elite leaders of native-place groups, ritual activities made it possible for social institutions to function effectively.

At the same time, the self-defense forces that native-place coalitions relied on to protect their ecological turf could also create serious conflict and instability. With each native-place coalition prepared to employ force to protect its ecological niche, competition over limited resources could easily escalate into violence. When feuding between opposing native-place groups took place, local officials frequently intervened to carry out mediation. But even when mutual defense gave way to mutual warfare, officials relied on the elite leaders of regional fishing lodges to enforce and maintain settlements, keep the peace, and prevent further disorder.

In every instance, social institutions coordinating the use of common-pool resources also functioned as instruments defining the distribution of wealth and relations of power.[83] By enforcing rules regulating the use of Zhoushan's fisheries, local elites secured access to profits derived from the marine environment.

Beginning in the late nineteenth century, fishing lodge organizations also gave officials a conduit for tapping into fiscal revenues collected from fishing enterprises. The liturgical role that the leaders of fishing lodges played in collecting taxes and maintaining social order garnered official backing for local arrangements governing the use of fishing grounds.[84] Local officials and elites had a shared interest in dispute resolution, since losses due to violent conflict over resources cut into commercial profits and tax revenues alike.

Localized disputes over fishing grounds proved less difficult to resolve than the struggles over hydrological systems investigated by Peter Perdue and Keith Schoppa, where large-scale problems of coordination involved entire regions and necessitated vigorous official action to protect public water resources against private encroachment. At the same time, the goals of these official interventions were different from those of the regulations that governed Zhoushan's fisheries. Perdue and Schoppa detail the fights that Qing officials waged against illegal dike building and polders to ensure long-term preservation of lakes and prevent future flood disasters.[85] By contrast, the rules that coordinated use of Zhoushan's fisheries during the Qing facilitated extraction of natural resources by minimizing violent disputes and protecting short-term profits. It should come as little surprise that keeping exploitation going was far easier than trying to hold it back.

The regulations governing Zhoushan's fisheries promoted human welfare by allocating access to scarce resources, maintaining profits by limiting conflict over fishing grounds. Fishermen in Zhoushan succeeded in overcoming the problems that could result when boats physically interfered with one another or did not efficiently allocate access to productive fishing spots. But unofficial arrangements did nothing at all to restrict environmentally damaging gear or keep catches at sustainable levels. Still, it should be pointed out that fishermen in China's Zhoushan region were not exceptional in this regard. The unpredictability of the marine environment makes it difficult to distinguish between the impact of human fishing activity and naturally occurring fluctuations in the abundance of fish stocks. Hence, institutional arrangements devised by fishing communities all over the world have typically proven more effective in settling conflicts over common-pool resources than in averting overexploitation.[86]

THREE

Developing the Ocean: Expansion and Reform, 1904–1929

The ecological dynamics that fueled the growth of the Zhoushan Archipelago's fishing industry under the Qing persisted and gained even greater momentum during the Republican period (1911–49). Interlocking economic and environmental changes amplified human demands on the marine ecosystem. During the early twentieth century, economic integration with regional marketing systems facilitated more effective exploitation of the ecological niche afforded by Zhoushan's fisheries. As the Zhoushan Archipelago's marine environment grew ever more closely tied to coastal markets, expanding credit networks connected migrant fishing enterprises with capital flowing from booming regional economic centers. Small-scale fishing enterprises took advantage of these financial relationships to exploit fish stocks more intensively, thereby heightening competition for limited resources. In this respect, the expansion of Zhoushan's fishing industry was linked to the growth of Shanghai, Ningbo, and other urban centers in the 1910s and 1920s, which deepened demand for fish products.

While the development of Zhoushan's fisheries continued apace from the nineteenth to the early twentieth century, this era also witnessed the emergence of new ways of understanding human relationships with the marine environment. From the final years of the Qing into the Republican period, reform-minded elites espoused means of managing China's fisheries sharply different from the arrangements used by regionally based fishing groups to exploit marine resources profitably. These initiatives reflected the modern Chinese state's larger goal of harnessing the power of science to obtain maximum production from nature. This particular conception of modern science rested on the belief that the environment was infinitely malleable and that, if proper methods were employed, humanity would not have to "beg for favors from nature" but would be able to command it and shape it at will.[1] Policies informed by modern science and undergirded by modern technology were to facilitate rational exploitation of nature and promote economic development.[2] These objectives epitomized the global project of modernity pursued in China and throughout the twentieth-century world.

Reform efforts initiated during the final years of the Qing also created a group of foreign-educated Chinese fishery specialists who would later be employed by China's Republican-period governments. According to these technocrats, modern science and technology could maximize production and meet the rising market demand for fish while averting overexploitation. Chinese fishery experts interpreted declining catches from inshore waters as a sign of fishers' inefficient and wasteful concentration on the same fishing grounds year after year. To remedy this problem, fishery experts called for investigations to discover unexploited fishing grounds, as well as promotion of aquaculture to sustain increases in output. Whereas native-place groups allocated access to limited ecological niches, modern technocrats believed that official interventions could increase production indefinitely by expanding the available resource base. To this end, the modern Chinese state endeavored to reshape the institutions and organizations that native-place groups employed to maintain their share of profits from Zhoushan's marine environment. Fishery experts called for the replacement of regional fishing lodges with state-directed fishing associations, which would act as the mechanism for enacting new rules and practices governing the use of resources. In the early years of the Republican period, inadequate funding and the Chinese state's limited administrative capacity prevented fishery agencies from carrying out such plans, but their legacy proved long lasting.

Zhoushan's Fishing Industry in the Early Twentieth Century

Patterns of demographic expansion, internal migration, and commercialization in Zhoushan's fisheries continued virtually uninterrupted from the Qing to the Republican period. Unfortunately, precise statistics on the Zhoushan Archipelago's population during the nineteenth century are not available. In 1900 the registered population of Dinghai county, the administrative unit encompassing most of the islands, stood at 78,271 households and 344,890 individuals. Dinghai county's population rose throughout the early twentieth century, reaching 87,753 households and 383,739 individuals in 1931.[3] This influx of people included large numbers of fishermen from coastal Zhejiang and Fujian who settled permanently on the islands to carry out fishing production.[4]

Yet, such figures do not fully convey the changes that took place in these years. Zhoushan's fisheries possessed a largely sojourning character into the early twentieth century. Fishermen converged on the islands when the fish arrived and returned to their native place after the fishing season. As the gazetteer of Daishan Island published in the 1920s put it, "Recently, most of the profits from Daishan's fisheries leak out and are lost, because most fishermen are migrants from outside and few local people engage in fishing."[5] Locals fished with drift-net boats (*liuwang chuan*), catching herring from late spring to summer, as well as a modest numbers of crab and sole in autumn, winter, and early spring. Their numbers paled, however, in comparison to the masses of fishermen who came for only part of the year. Annual population cycles coincided with the migrations of fish species to the archipelago to spawn. The population of Daishan Island's two main fishing ports swelled by over 20,000 with the influx of sojourning fishermen and merchants from Taizhou, Ningbo, and coastal areas of Fenghua and Xiangshan counties, as well as other islands in the Zhoushan Archipelago at the height of the large yellow croaker runs from early May to late June.[6]

During the early twentieth century, Dongshajiao, the main fishing port on Daishan Island, showed the characteristic signs of a sojourning male population, "Every fishing season there is a forest of sails and masts. Brothels and gambling houses open up to take advantage of this demand, and itinerant traders and peddlers avail themselves of the chance to seek petty profits."[7] Trade flourished in Dongshajiao during fishing season, and shops in Gaoting, Daishan's other main market town, relied on sojourning fishermen for most of their business.[8]

The Dynamics of Economic Integration

Economic trends in the early twentieth century linked Zhoushan's marine ecosystem more closely with demand from major urban consumption centers, facilitating the expansion of Zhoushan's fishing industry and placing heightened pressures on the natural environment. The period from the 1910s to the mid-1920s saw an upsurge in China's modern economic sector, as World War I temporarily eliminated European competition.[9] This boom redounded to the benefit of fishing enterprises in Zhoushan. Economic integration with Shanghai and other urban centers in the Lower Yangzi and Southeast Coast regions gave small-scale producers the potential to reap greater profits from fish stocks and enhanced their ability to exploit the marine environment. Expanding networks of credit acted as a driving force behind ecological change, making it possible for undercapitalized fishing enterprises to accelerate the extraction of resources.

The coastal peripheries of the Lower Yangzi and Southeast Coast macroregions were the origin of many fishermen who migrated to the Zhoushan Archipelago. The macroregions' cores were the source of the trade that fueled fishermen's livelihood. During the large yellow croaker runs each spring, passenger boats brought large numbers of sojourning merchants carrying goods from Ningbo, Wenzhou, Xiangshan, Fenghua, and other mainland locales. Cargo boats arrived with a wide array of supplies and left filled with fish products.[10] The advent of steamship navigation between fishing ports in the Zhoushan islands and commercial centers like Shanghai, Ningbo, Haimen, Wenzhou, and Fuzhou during the Republican period drew the islands into a growing web of urban markets on the mainland.[11] The development of Zhoushan's fisheries depended on increased demand for fish in these consumption centers, which in turn derived from economic growth in the Yangzi Delta and Southeast Coast during the Republican period.[12]

In the first three decades of the twentieth century, the financial networks that funded Zhoushan's fishing industry also grew more connected with major economic centers in the Lower Yangzi. Native banks (*qianzhuang*) with offices in Ningbo or Shanghai and Zhoushan's fishing ports acted as financial intermediaries for all business transactions between fishermen, ice boats, fish brokers, and wholesalers.[13] As commercial fishery production expanded, the number of native banks in the islands rose from one in 1908 to more than thirty in 1932.[14]

These financial institutions gave undercapitalized fishing enterprises a source of credit when they needed it most. Demand for credit peaked in early January with the onset of fishing season for hairtail. Like Zhoushan's other commercially important fish species, hairtail migrated from deep, offshore waters to coastal spawning grounds near the islands during the spring and early summer. However, the main fishing season for hairtail started when the fish made their annual migration from the Shengsi Islands to waters off the southern coast of Zhejiang each winter.[15] As one commentator observed during the 1930s, this fishing season brought major financial demands.

When hairtail fishing begins, all fishing boats need cash, and fish merchants strive to underwrite them. When the end of the fishing season arrives, all fishing boats have to repay their debts, so they must settle fishermen's pay and the like. Therefore, this period brings about a state of financial disarray. Some turn to high-interest loans, and lawsuits arise because of lack of funding. For fish brokers this is the period when credit is most essential.[16]

Fishing boats acquired much of their capital in the form of loans from fish brokers. Because most fish brokers had few capital reserves of their own, they relied on native banks for the credit needed to carry out financial exchanges with fishing boats.[17] The Yikang Native Bank (Yikang qianzhuang) in Wenzhou's port of Kanmen, which had a fish broker as one of its main shareholders, made loans mainly to fishermen who needed to repair their boats and purchase new gear. Every year during the summer fishing season, the bank's managers traveled to the port of Shenjiamen, where they set up a temporary office to collect repayment on loans made to fishing enterprises and remit these funds back to their native region.[18]

Much of the capital used by native banks on Daishan Island came from the money that shops in Shanghai, Hangzhou, and Shaoxing deposited to purchase the catch at the start of fishing season. Native banks in Daishan and other ports used this money to extend credit to fish brokerages and fish-processing enterprises, which loaned funds to fishing boats.[19] Native banks from Ningbo likewise came to Shenjiamen during fishing season to provide badly needed credit to fishing enterprises. Native banks in Shenjiamen and other ports set interest rates according to Ningbo's financial market, with a markup of 5 to 10 percent depending on fluctuations in silver prices.[20] These financial relationships reflected Zhoushan's growing integration with Shanghai, Ningbo, and other economic centers in the Lower Yangzi. Pawnshops, which flourished in fishing centers such as Shenjiamen, Daishan, and the Shengsi Islands in the late nineteenth and early twentieth centuries, acted as additional sources of credit.[21] The pawnshop business paralleled the timing of fish runs, with fishermen pawning goods prior to the spring and winter fishing seasons and redeeming them in the fall.[22]

In Zhoushan's fisheries, financial institutions were conduits for the pattern of commercial integration that gave core economic regions access to natural resources located in their peripheries.[23] A Japanese investigation of Zhejiang's fisheries from the mid-1920s summed up the financial relationships between coastal fishing regions and marketing centers like Ningbo and Shanghai in the following terms:

Most fishermen engage half in farming and half in fishing, establishing villages in coastal regions. They are resigned to a meager life, and most of their capital comes from fish merchants in cities. . . . As compensation for advance loans (actually, merchants do not provide cash but mainly loan miscellaneous supplies that they procure themselves), access to the catch is put in the hands of fish merchants. They control fishing people in perpetuity, making profits upon profits, and leave almost no surplus for fishing people's livelihood. Therefore, fishermen lack adequate capital and their places of residence remain peripheries.[24]

Overlooked in this analysis of the connections between fishing regions and economic centers was the impact that financial relationships had on the natural environment. In Zhoushan's fishing industry, credit from economic centers permeated the marketing relationships that transformed natural resources into commodities. Fish merchants acted as the nexus that connected labor from coastal fishing villages and investment capital from economic centers, making intensified exploitation of fish stocks possible. Connections between Zhoushan's fisheries and urban centers in the Lower Yangzi—China's most economically advanced macroregion—made the ecological effect of these financial relationships particularly acute. In less-developed parts of China with lower levels of demand density, commercial activity, and capital investment, financial institutions would not have been such an important factor in environmental change.

With greater access to credit in the early twentieth century, fish brokers in the Zhoushan region could invest profits in expanding their business operations and offer loans to a larger numbers of fishing boats. Daishan Island's native banks, for instance, borrowed most of their capital from larger native banks in Ningbo. Native banks in Daishan used this capital to offer credit to fish brokers and fish-processing enterprises, which in turn loaned funds to fishing boats.[25] On one hand, expanding networks of credit gave undercapitalized business enterprises a way to cash in on natural resources more effectively. On the other hand, reliance on credit forced fishing enterprises to bring in more and more fish to pay off their debts. These financial relationships thus fueled the growth of the fishing industry but also intensified pressures on the marine environment.

Conflict and Conciliation

As Richard C. Hoffmann has pointed out, heightened demands on limited or diminishing resources often trigger contests for control.[26] As migration and economic integration put rising pressure on Zhoushan's fisheries in the early twentieth century, intensified competition gave rise to tensions among fishermen's native-place groups. During the late 1920s, observers noted that in the waters around the Shengsi Islands, "The fishing rights are well-established by custom." Although the Shengsi Islands fell under the political jurisdiction of Jiangsu province, most of the fishermen came from Zhejiang and Fujian, and only a few boats from the Chongming Delta fished near the coast. The threat of violent reprisals backed up each regional group's proprietary claims to these fishing territories. "By custom no fishermen from the Central or Northern section [of Jiangsu] is allowed to come south of the Yangtze Estuary. Disregard of the custom always leads to bloody feuds and destruction of the trespassing vessel."[27]

As competition sharpened for Zhoushan's fishing grounds, many fishing lodges formed militias to defend their claims. Funds that lodges collected from fishing boats met the cost of supporting mutual-defense organizations. As the Yongan Lodge's 1915 regulations stated, "The Shengshan Sea is located on the Jiangsu-

Zhejiang border, and it is truly a place of refuge for pirates. For the past ten years this group's fishing households have paid fees for their own defensive convoy."[28] Protection against piracy justified these elite-led paramilitary organizations, but they also gave native-place coalitions a means of protecting their ecological niche against competitors.[29]

Along China's southeast coast, the line between piracy and fishing was always a fine one. Fishermen took part in occasional acts of piracy, and pirates targeted fishermen and their catch.[30] In a petition from the early 1930s, the Renhe Lodge and several other fishing organizations in Shenjiamen asked the local government to help them crack down on "unlawful fishermen" who rammed other boats, made off with their nets, and took them for their own as pirates did.[31] To fishermen, the only difference between rival fishing boats and pirates was that the former took fish before they were caught, and the latter took fish from their nets. Defense against piracy thus overlapped with the efforts that lodges undertook to keep competing fishing groups out of their ecological turf.

As competition for common resources intensified during the Republican period, local elites continued to take primary responsibility for diffusing violent conflicts over fishing grounds. Long-standing tensions existed between fishermen from Tongzhao village in Fenghua and Dongmen village in Xiangshan county, all of whom fished with two-anchored stow nets in the waters off Daishan Island during the summer and autumn. During the late Qing, the two groups engaged in a violent feud, from which the Fenghua fishermen emerged victorious. In the 1890s, the groups reached an agreement that divided Hengjie, the fish market in the port of Dongshajiao, into two separate spheres of influence. From then on, fishermen suffered brutal beatings whenever they crossed into the other native-place group's territory. When arguments broke out, the Xiangshan fishermen's Taihe Lodge and the Fenghua group's Yian and Yihe lodges handled the situation, and local officials did not get involved.[32]

Despite the lodges' efforts to maintain order, in 1921 feuding once again erupted between the Fenghua and Xiangshan groups. During a gambling dispute, an armed mob of fishermen from Xiangshan killed a fisherman from Fenghua. After notifying local officials about the incident, the Fenghua group began to accumulate a large cache of weapons. Their Xiangshan rivals dug trenches and built fortifications to oppose them. Stray bullets from the "small war" that ensued injured many innocent bystanders and forced shops in the port to close. Merchants in Dongshajiao, whose business suffered as a result of the dispute, held an emergency council urging the two groups to take their confrontation outside the town. The merchants also appealed to local officials to stop the conflict. On the following day, the police chief and the commander of the local military garrison met with the leaders of the Fenghua and Xiangshan fishing lodges to reach a truce.

After the fighting broke out, the prominent Fenghua gentryman Zhuang Songfu came to Daishan Island and took over as director of Fenghua fishermen's Xifeng Lodge to assist in negotiations.[33] Previously, Zhuang had worked closely with the Xifeng Lodge's director Jiang Beiming setting up educational institutions in Fenghua and in running a reform-oriented publishing house in Shanghai. Despite Zhuang's presence, talks between the fishing groups fell apart, and they began preparing for a decisive battle. The local police and military forces appealed to county officials to come to Daishan and handle the situation. After another day of fierce fighting, the Dinghai county magistrate Zhang Yin arrived and ended the feud.[34]

Yet even after officials intervened, local elites took responsibility for formulating a settlement. To resolve hostilities once and for all, Fenghua and Xiangshan fishing lodge leaders enacted a treaty forswearing armed feuding. Officials granted their approval to this agreement after leaders from the fishing groups had signed the pact. Religious associations in Daishan again functioned as the mechanism for putting these rules into effect. In 1923 the treaty was posted in an inscription at Dongshajiao's God of

Wealth Temple (Caishen dian). Its terms spelled out regulations for resolving disputes over fishing territories. Fishing lodges handled most of these conflicts, but the organizations could turn to government officials for support if disagreements got out of hand.

The treaty required fishing fleets to stay in their mooring areas and refrain from entering the other's territory. In the event of disagreements, fishermen were to report the incident to their lodge organization for resolution. Lodge directors threatened to punish fishermen who took part in feuding. Leaders of fishing lodges also undertook to stop crew members from creating disturbances in "wineshops, teahouses, or other places" ashore. Fishermen who took part in such incidents would be handed over to their fishing lodge, which would send them to officials for punishment if necessary. The local militia and police would use force to disperse either fishing group if they assembled a crowd of more than three people.[35] If additional feuding took place, local officials would punish all participants in the conflict and penalize each lodge's "brokers and fishing headmen" (*hanghu yushou*) for their negligence. Each lodge appointed an inspector (*xuncha*) to follow the naval patrol ships stationed at Daishan during fishing season and report illegal activities. Local officials would punish fishermen along with their lodge leaders if they tried to take vengeance by making accusations against their rivals.[36] Even as competition for limited resources intensified, patterns of dispute resolution persisted from the Qing into the Republican period. Despite the political upheavals that shook China during the first two decades of the twentieth century, local authorities continued to rely on elite groups to achieve the shared objective of preventing the loss of profits to violent conflict.

Zhang Jian and the Beginnings of Modern Fishery Management

In the first decade of the twentieth century, Japan rapidly became influential in China as a model of modernity and a filter for Western influences. So it was in the fishing industry. As Zhoushan's fishing industry grew and competition for resources mounted during the first three decades of the twentieth century, reform-minded Chinese officials began to formulate plans to remake the fishing industry according to a modernist vision of scientific fishery development that originated in Europe and the United States. Chinese reformers gained their exposure to these global environmental discourses by way of Japan, which had refashioned Western knowledge and expertise to fit its emerging vision of modernity in the late nineteenth century. During the Meiji period (1868–1912), Japanese fishery experts had taken a keen interest in the aquaculture, fish-processing, and manufacturing techniques they learned from foreign advisors and observed during visits to the United States and Europe. In Meiji Japan these forms of fishery management—much like scientific agriculture—were framed as thoroughly modern and progressive.[37] During the first decade of the twentieth century, these modernist environmental approaches began to exert a profound influence in China.

Fishery production may have grown steadily along traditional lines during the late nineteenth century, but for these reformers an awareness of the vulnerability of China's fishing industry to foreign competition seemed to necessitate far-reaching changes. In the final decade of the Qing dynasty, the central government pursued reforms, known as the New Policies (*xinzheng*), inspired by Western and Japanese models in a last-ditch move to shore up its waning authority. As part of the New Policies, the official and entrepreneur Zhang Jian launched the earliest call to modernize China's fisheries according to the blueprint provided by Japan's fishing industry.

Zhang Jian first advocated changes in China's fisheries after a visit to Japan in 1903, where he observed the advances in Japan's fishing and shipping industries that had taken place during the late nineteenth century.[38] In a petition to the Ministry of Commerce, Zhang Jian drew a direct parallel between the development of a nation's fisheries and its maritime strength:

Sea power and fishing territories are one and the same. Sea power is of the nation; fishing territories are of the people. If fishing territories are

not clarified, one cannot establish sea power; if sea power is not expanded, one cannot protect fishing territories. They maintain each
other, and it is like this for all nations. China has never had a fishing
administration, and its fisheries lack organization. The oceans that its
fishing boats reach are in scattered regions. Wei Yuan's *Illustrated Trea-
tise on the Maritime Countries* (*Haiguo tuzhi*) and other books are far inferior to the detailed records of the British admiralty charts. As to the
doctrine of sea power, most literati cannot converse on it thoroughly.
At this time, bans on maritime trade have been lifted, the whole world
is converging, and all countries are striving to expand their sea power. If
we do not formulate a plan before it is too late, our fishing territories
will be violated because of our passivity, and our sea power will be in
dire straits because of our concessions.[39]

According to Zhang Jian, fortifying the nation's external boundaries required expansion and consolidation of fishing rights. Failure to do so had made the nation's maritime frontier vulnerable
to external threats. China's lack of attention to maintaining its
fishing rights explained its inability to stand up to foreign aggression since the mid-nineteenth century. Exerting control over marine fishing grounds was a crucial way to safeguard China's sovereignty and promote national wealth and power. For this reason,
Zhang Jian called for assertive measures to block foreign efforts
to "encroach on our country's sea power and plunder our people's
profits from fishing."[40]

    To prevent losing fishing grounds to foreign nations, Zhang
Jian proposed establishing a fishing company in Zhejiang and
Jiangsu, with similar enterprises in Fujian and Guangzhou to follow. The company would introduce modern fishing technology by
purchasing a steam-driven vessel to patrol fishing grounds and
prevent foreign encroachments. With private capital not forth-
coming, Zhang obtained a government loan of 50,000 taels to
purchase a steamship (later rechristened the *Fuhai*) from a failed
German fishing enterprise in the Shandong port of Qingdao. In
1905, the Jiangsu-Zhejiang Fishing Company (Jiang-Zhe yuye
gongsi) opened in Shanghai with Zhang Jian as its manager.[41] The
Fishing Company's steam-powered trawler was to compete with

foreign fishing vessels on the high seas, while ordinary boats fished inshore waters around the Zhoushan islands. Since the two types of boats fished in entirely different areas, it was believed, the mechanized vessel would not create losses for small-scale fishermen. The Jiangsu-Zhejiang Fishing Company's regulations stated:

All the places where two-anchored stow-net boats fish are close to islands. Since the steam trawler must avoid reefs, it will definitely not interfere with them. In places where there are drift-net boats, the steamship will also give way and will definitely not occupy their fishing grounds. The places where every other kind of boat casts their nets and their markets will also remain the same and will not be disturbed.[42]

Once in place, this arrangement was to safeguard Chinese fishermen's profits by keeping foreign vessels out of fishing grounds close to shore. At the same time, the Jiangsu-Zhejiang Fishing Company's steam-powered trawler would extend China's fishing territories by opening up fishing grounds on the high seas.[43]

As part of the Jiangsu-Zhejiang Fishing Company, Zhang called for a fishing association (*yuhui*) with branches in each of China's coastal provinces. Fishing militias and mutual-security units (*baojia*) were to change their names and form these new organizations. Zhang expected leaders selected from among "elders whom the masses obey" to compile registers of the fishing population in each area. From time to time, the Fishing Association would depute a steamship to inspect its local branches.[44] As Zhang stated, "If there are no branch associations, there will be no way of clarifying household registration or investigating and protecting the thousands of fishing households that are dispersed in places with numerous streams and islands."[45] The Jiangsu-Zhejiang Fishing Company's steamship would defend boats that joined the Fishing Association against piracy. To this end, the vessel was outfitted with artillery, guns, and swords.[46] Enjoying this protection, fishermen would appreciate the Fishing Association and would not be "taken in by foreigners' enticements."[47]

At the same time, Zhang looked to the Fishing Association to seize a larger share of profits created by the commercialization of

Zhoushan's fisheries in the form of customs duties levied on fish marketed in Shanghai. Many of the ice boats that carried fresh fish from Zhoushan's fishing grounds to Shanghai had taken to purchasing French flags from Chinese merchants who had connections with foreign interests in the city. This scheme made it possible to avoid payment of *lijin* transit duties and custom levies, as well as the "squeeze" demanded by the personnel in charge of collecting them. To consolidate the Qing government's control over these lost revenues, the Jiangsu-Zhejiang Fishing Association (Jiang-Zhe yuhui) collected fees at a rate lower than the amount merchants charged to give ice boats foreign flags. Responsibility for collecting these revenues fell to fish brokers in Shanghai. Headmen (*zhushou*) from each native-place group made a list of the registration numbers assigned to each ice boat for the Jiangsu-Zhejiang Fishing Company's reference and took responsibility for punishing those caught trying to avoid these fees.[48]

For a time, revenues from the Fishing Association enabled the Jiangsu-Zhejiang Fishing Company's trawler to continue its operations even though it never turned a profit. But by the 1920s, ongoing financial losses led to the Fishing Company's dissolution. After the company ceased to exist, the Fishing Association took over the steamship *Fuhai,* using it to protect ice boats from pirates. Every fishing season, the Fishing Association's patrol boats traveled to the Zhoushan Archipelago's waters to safeguard the silver used by fishermen and ice boats in their commercial transactions. As a result, the Fishing Association's steamships turned into the nucleus of the "market at sea" in Zhoushan's fishing grounds.[49] During the annual yellow croaker and hairtail runs, the Dunhe Lodge formed by Shanghai's frozen fish merchants joined with the lodges of ice boats from Ningbo and Taizhou to hire coast guard officers to man their convoys. Ge Liquan, the director of the Taizhou Lodge, oversaw these mutual-defense activities.[50] From the late Qing into the Republican period, fishing lodges in the Zhoushan region thus retained firm control over tax collection and local defense.

Modern Fishery Management in the Early Republican Period

Zhang Jian's plans for modernizing China's fishing industry included fishery research organs modeled on those in North America, Western Europe, and especially Japan.[51] This modern discipline of fishery studies concerned the effective utilization of aquatic animal and plant life. One of the earliest fishery studies textbooks, translated from Japanese into Chinese in 1911, explained that even though people did not exploit all these natural resources, all of them had potential uses. If any were discarded, it was not because they were useless. Rather, it was because "research on the techniques for utilizing them is not refined." Making efficient use of marine resources required knowledge drawn from fields as diverse as zoology, botany, physics, chemistry, geography, oceanography, and economics.[52]

To disseminate knowledge of these scientific techniques, Zhang Jian's original proposal for the Jiangsu-Zhejiang Fishing Company called for setting up a fishery school at the company's headquarters in Wusong.[53] In 1912, a graduate of the Tokyo Fishery Institute named Zhang Liu set up the Jiangsu Provincial Fishery School with support from Zhang Jian.[54] During the next two decades, several other fishery schools and research centers opened in China's coastal provinces. Throughout the Republican period, nearly all the fishery experts who held faculty positions in these educational institutes had been trained in Japan. As a newspaper report from the 1920s stated, "Most instructors in fishery schools in this country have studied abroad at Japan's Tokyo Fishery Institute or have engaged in observation and study in Japan."[55] Drawing on this training, China's first generation of fishery experts introduced new ways of thinking about human interactions with the marine environment.

In the eyes of Chinese fishery experts, the biggest problem with China's fishing industry was its inability to keep up with surging domestic demand. Imports of marine products, the bulk of which came from Japan, outpaced exports during the late nineteenth and early twentieth century.[56] Of course, comparing China's import and export figures overlooked the huge amount of fish products marketed in China for domestic consumption. In Shanghai, which was by far China's largest fish consumption center, domestically produced fish vastly outweighed foreign imports from the 1910s into the 1930s.[57] Nevertheless, this economic situation did not figure into the nationalist rhetoric of Chinese fishery experts. In an article published in the journal of the Jiangsu Provincial Fishery School in 1918, Zhang Liu explained that China's fishing industry had experienced considerable progress in recent years, but remained at the "sprouts" (*mengya*) stage. The country still relied far too heavily on imports: "In a word, our country's demand for fishery products increases with each day, while foreign imports also grow in quantity and our rights to profits slip away (*li-quan waiyi*)."[58]

The solution to what fishery experts saw as the backwardness of China's fishing industry lay in modern science and technology. As one fishery specialist succinctly stated in the early 1920s, "Scientific methods should be used to solve fishery problems." More concretely, scientific methods meant improving fishing gear, refining processing and storage, and promoting aquaculture. To teach the fishing population about these new methods, China's government needed to set up fishery research stations staffed by "scientific specialists."[59] Chinese fishery experts found inspiration for these training centers in the research and education institutes that had contributed to fishery development in Japan during the Meiji era by disseminating improvements in gear and processing methods.[60] These Japanese fishery schools, of course, were precisely where Chinese fishery experts had received their education.

Efforts to set up fishery studies institutes in the Zhoushan Archipelago began with Li Shixiang, a native of Jiangsu's Chongming county who had studied at the Tokyo Fishery Institute from 1907 to 1911.[61] In 1918, China's central government sent Li Shixiang to open the Dinghai Fishery Training Institute on Zhoushan Island.[62] Unfortunately, Li Shixiang's early effort to reform

Zhoushan's fishing industry ended in disappointment due to in-adequate funding.[63]

The following year, China's central government deputed Wang Wentai, another Japanese-educated fishery expert from Jiangsu, to set up a fishery-training institute in the northern Jiangsu port of Haizhou. As with the Dinghai Fishery Training Institute, lack of funds limited Wang Wentai's endeavor.[64] Nevertheless, his detailed outline for the Haizhou Fishery Training Institute illustrates how Chinese fishery experts perceived human interactions with the marine environment. One of Wang's early reports on Haizhou's coastal fishing grounds captured the characteristics of marine resources under strain from human exploitation. As Wang explained, fishermen's harvesting of the same inshore fishing grounds year after year had led to localized depletion:

The output of fish species in coastal seas has decreased in recent years, and their size has also become increasingly thin and small. . . . This is because fishermen take advantage of fishes' migration patterns and overfish when they come to coastal waters to spawn. They do not know about expanding the scope of marine fishing grounds. Unfortunately, if this continues, it will result in the extinction of migratory fish species in coastal waters, and this is regrettable.[65]

Thanks to the conservatism and ignorance of China's fishing population, Wang maintained, the fishing industry was "still in a half-developed epoch, and as a consequence it cannot make full use of nature's great benefits." To remedy this situation, fishery research organs had to assist fishermen by opening new fishing grounds. At the same time, innovations in aquaculture would replenish declining inshore fish stocks. Without positive steps to ensure the abundance of marine resources, the introduction of new fishing techniques would aggravate the problem of overfishing and "searching for profits would lead to losses instead."[66]

Wang Wentai and other Chinese fishery experts were aware that overfishing could deplete specific fishing grounds and even lead to the extinction of certain species. Yet this awareness existed alongside a belief that modern science and technology could make exploitation of natural resources optimally efficient, perpetually expanding production. Achieving this end required regulations that prohibited fishing methods and gear that impeded reproduction of stocks by capturing young fish before they grew to full maturity. According to a fishery studies textbook published in 1919, "Even though fishery products propagate extremely rapidly, to sustain their development forever there must also be artificial protection to assist their natural reproduction."[67] With advances in aquaculture, modern technology also made it possible to plant fish species in areas that did not yet produce them.[68] If measures were taken to protect the reproduction of fish species, "they can be inexhaustible and nature's abundant resources may be secured forever."[69]

Fishery experts also stressed the importance of scientific research to identify new fishing grounds and point boats to the areas where fish were most plentiful. Government agencies needed to introduce new technologies that enabled fishermen to locate and exploit productive fishing grounds. These investigations, it was believed, would relieve pressure on overexploited fish stocks and give them the chance to replenish.[70] Once again, this logic grew out of a familiarity with the development of Japan's fishing industry that Chinese fishery experts gained during their studies abroad. During the Meiji era, Japanese fishery programs overcame a perceived stagnation in fishery production caused by concentration on overexploited inshore waters. To remedy this problem, the Meiji government successfully expanded the territorial scope of fishing grounds and encouraged the use of more efficient technologies.[71] To achieve these goals, the Japanese government enacted legislation in the 1890s that offered financial incentives to boats capable of fishing in offshore waters.[72] In similar fashion, Chinese fishery experts believed that China needed to follow Japan's example and take action to overcome declining returns from inshore fishing grounds.[73] Following these recommendations, the Chinese government enacted Regulations for the Encouragement of Fishing on the High Seas (*Gonghai yuye jiangli tiaoli*) in 1914.[74] In contrast to Meiji Japan, however, the Chinese state did not actually implement these fishery development measures. Few subsidies were distributed to vessels capable of fishing on the high seas, and the policy had no real effect.[75]

Rationalizing Fishing Organizations

To Chinese fishery experts like Wang Wentai, disseminating knowledge about the most efficient ways of using marine resources also required transforming the fishermen's social organizations. In Wang's estimation, Haizhou's dispersed fishing population did not communicate or share information with one another and lacked adequate collective associations. Fishermen stuck to their old types of gear and had no way to implement improvements. Yet Wang believed the situation differed in Zhejiang, where fishing lodges already exerted discipline over the fishing population: "Zhejiang province's fishing lodges are thorough in rectifying improper behavior within their trade. If fishermen are not moral and just toward other fishermen and their investors (*zibenzhu*), the lodge can punish them without going through an official examination. Fishing lodges can even mete out capital punishment, and fishermen peacefully accept it." Still, Zhejiang's fishing lodges paled in comparison to the cooperative fishing associations (Ch. *yuye zuhe*, J. *gyogyō kumiai*) that Japanese colonial authorities had set up in territories they occupied in Shandong and Manchuria. Wang explained that the Japanese fishing organizations "are similar in character to our country's fishing lodges, only their regulations are more complete, the scope of their activities is vast, and their results are more far-reaching."[76] If the Chinese government encouraged fishermen to organize same-trade associations (*gonghui*) like those set up by the Japanese, fishermen would finally be able to further their mutual interests and develop the industry. Hence, local officials needed to direct the formation of new fishing organizations from above, while gentry-managers (*shendong*) advocated them from below.[77]

Wang Wentai presented a blueprint for these organizations in a report on Japanese fishery activities in the Shandong port of Qingdao, which he forwarded to the central government in 1922. Following their seizure of the German leasehold in Qingdao after the outbreak of World War I, the Japanese implemented regulations in 1916 that required all fishing enterprises to register with the city's military command. The Japanese also required all fishermen to join a fishing association, the tasks of which were to "reform and develop the fishing industry, protect the reproduction of fish species, correct the harmful trade practices of cooperative members, and promote their mutual benefits."[78] Besides offering incentives to fishing boats that adopted more efficient gear and processing techniques, the association gave fishing enterprises access to credit and affordable supplies. The cooperative association worked in conjunction with a Japanese-run fish market that monopolized the right to sell fish in the port.[79] Mechanized refrigeration facilities made it possible for market facilities to store fish for sale during periods of peak demand regardless of seasonal fluctuations in supply. As Wang Wentai pointed out, these arrangements not only stabilized prices but also safeguarded credit, expanded sales channels, and facilitated financial transactions between producers and merchants.[80] In this way, the Japanese streamlined the relations between nature, producers, and the market that turned fish into commodities. The increased production facilitated by these marketing arrangements was such that "when compared with the fish brokerage system in our country's various regions, their advantages and disadvantages cannot be spoken of on the same day."[81]

By the 1920s, these measures had made Qingdao into the center for Japanese fishing activities in the Bohai and Yellow seas off the northern coast of China. The Qingdao-based Japanese fleet sold much of the fish it caught in these waters on markets in north China.[82] Wang Wentai claimed that Japanese competition stifled China's fishing industry by driving down prices and forcing out domestic producers. According to Wang, overfishing by Japanese trawlers in the Yellow Sea also reduced the Chinese fishing fleet's catches.[83] Mechanized Japanese boats started to fish in the Yellow Sea in greater numbers after the Japanese government restricted steam trawlers to waters west of 130 degrees longitude (roughly the East China and Yellow seas) in 1912.[84] (These policies and the impact of Japanese fishing activities off the coast of China during the 1920s are analyzed in Chapter 4.)

Faced with economic difficulties caused by Japanese competition, Wang Wentai claimed, many impoverished fishermen had turned to piracy, which presented even greater obstacles to the Chinese fishing industry's development. Just as important, Wang pointed out, the Japanese control of Qingdao's fish market deprived the Chinese government of a lucrative source of tax revenues.[85] Foreign incursions on fishing grounds off the coast of China were thus a direct threat to national sovereignty:

The Japanese have operated there [in Qingdao] for eight years, impeding our country's sea power and profits from fishing. Not one of the policies that their government puts into effect does not decimate our fishing industry and exploit our fishermen, sucking our blood and sweat for their own progress, and using pirates to cause chaos on the seas. Their impact reaches far beyond Qingdao.[86]

To counter this threat, Wang called for China's central government to carry out measures like the ones the Japanese had put in place. Wang also urged the central government to take steps to reclaim Qingdao's fish market to consolidate China's fishing rights in the Yellow Sea and regain this source of tax revenues.[87]

Other Chinese fishery experts echoed Wang in pushing reform of fishing organizations and marketing arrangements as the first step toward rejuvenating China's fishing industry. When the Chinese government set up the Dinghai Fishery Training Institute in 1918, it announced its intention to gather information on credit and marketing associations like those found in foreign countries.[88] In 1921 the Ministry of Agriculture and Commerce revealed an elaborate agenda for revitalizing China's fishing industry, with plans to set up marketing facilities that would collect fees to fund education and relief for the fishing population. Same-trade associations would also promote fishermen's mutual interests.[89] Motivations for this scheme derived from several sources. The Japanese reorganization of Qingdao's fisheries had acted as an "outside stimulus" for the move to develop China's fishing industry.[90] Additional impetus for reform of the fishing industry came from fishery experts like Wang Wentai and Li Shixiang, who pressed the Chinese government to implement their fishery-development proposals.[91] In 1924 the Ministry of Agriculture and Commerce undertook a survey of coastal areas to find locations suitable for the construction of modern fishing harbors. The ports were to be equipped with centralized marketing facilities, transportation, and refrigeration equipment.[92] Similar ideas made their way into the *Industrial Plan* (*Shiye jihua*) written by Sun Yat-sen (Sun Zhongshan), which called for the construction of modern fishing harbors along China's coast, including one on Changtushan Island in the Zhoushan Archipelago.[93]

Streamlining the marketing system went hand in hand with improving fishing organizations. State-run fish markets would facilitate financial transactions by setting up fishery banks to provide credit to producers through fishing associations.[94] The associations would also organize credit, purchasing, and marketing cooperatives to give producers low-interest loans and affordable supplies, decrease production costs, and bring fishermen a higher price for the catch.[95] In all these respects, fishing associations resembled the agricultural cooperatives that many reformers expected to solve China's rural economic problems during the Republican period.[96]

These goals still informed fishery policy after 1927, when the Nationalist party-state came to power with the dual mission of defending China's sovereignty and building a powerful national economy. Jin Zhao, a Japanese-trained fishery expert from Zhejiang, emphasized the rectification (*zhengdun*) of fishing organizations in his proposals for improving Zhejiang's fisheries during the early 1930s.[97] In his plans for the fishing industry, Jin gave top priority to fishing associations:

The main reason why the old fishing industry has failed is that fishermen do not have collective organizations. They lack reliable capital and suffer exploitation from fish brokers and fish merchants, which puts fishermen's livelihoods in jeopardy. Therefore, the first step in the plan should be to straighten out and improve all of fishermen's old evil habits and to organize fishing cooperatives.[98]

With these new associations in place, fish brokers and fish merchants would no longer have the power to "monopolize" the market for fish. All economic transactions would be made "free and independent." Researchers needed to investigate new fishing grounds and inform "dull-witted" fishermen who "do not understand the movements of schools of fish" about their location.[99] Jin acknowledged that larger catches had the potential to reduce even the most abundant fishing grounds. But he was confident that expansion into new, more productive fishing grounds along with research on aquaculture could sustain increased production indefinitely.[100]

Referring to Sun's *Industrial Plan*, Jin Zhao called for state-run markets in Changtushan and eight other modern fishing harbors along Zhejiang's coast. These markets would operate as "big fish brokerages" equipped with mechanized refrigeration equipment and other up-to-date facilities.[101] In place of fish brokers, Jin asked the provincial government to organize cooperative banks to extend low-interest loans to fishermen.[102]

With these new organizations and new technology, natural resources can be opened up and used to enrich the people. These so-called new organizations and their socialization (*shehuihua*) will give scattered fishermen systematized organization. . . . So-called new technology is scientific. Science should be used to improve old fishing techniques and processing methods, and aquaculture methods should be used to rationalize production.

The new sources of wealth gained by increasing fishery output would stabilize social order and fortify the nation against external aggression.[103] With the reform of fishing organizations and the application of modern science, state intervention would make the use of marine resources more efficient and productive.

For Chinese fishery experts, the goal was to unleash nature's potential by eliminating waste and rationalizing exploitation of the marine environment. Their developmentalist vision derived from a faith that centralized, scientific management could generate unlimited output from finite resources. Fishery experts believed that if marketing reforms were carried out in tandem with measures to expand the overall resource base through aquaculture and the opening up of new fishing grounds, their plans could make the ocean endlessly productive. Clearly, these modernist goals differed fundamentally from the social institutions that particularistic, regionally based fishing groups relied on to maintain their share of limited common resources.

Republican Period Fishery Legislation, 1922–29

The reform programs advanced by Chinese fishery specialists required the creation of new institutions governing the use of fish stocks and organizations capable of putting these rules into effect. Throughout the 1920s, China's modern state sought to "encourage fishermen to form associations to carry out mutual aid, deal with fish merchants to set them on the right track, and subject same-trade associations that have already been established to official supervision in accordance with the law."[104] Realizing this goal required creating state-directed fishing organizations inspired by Japanese precedents. These efforts began with the Temporary Fishing Association Regulations (*Yuhui zanxing zhangcheng*) of 1922 and the Rules for the Implementation of the Temporary Fishing Association Regulations (*Yuhui zanxing zhangcheng shixing xize*) of 1923.[105] The Nationalist regime built on these earlier pieces of legislation in 1929 with the Fishing Association Law (*Yuhui fa*).[106] The Fishing Association Law, like other Nationalist economic legislation, revealed a desire to both promote and control economic enterprise, while strengthening state power.[107]

The Fishing Association Law spelled out detailed guidelines for the formation of collective organizations dedicated to the improvement and development of China's fisheries. The law called for each county to set up one fishing association, but there could also be branch associations (*fenhui*) in "thriving fishing areas" located over forty *li* from the main association. Fishing associations were to encourage fishermen to adopt more efficient technologies and organize cooperatives for credit, production, and marketing. The law also made the associations responsible for mediating fishing disputes. The new organizations would thus take over the tasks previously performed by fishing groups' native-place lodges.

In addition to a commitment to developing China's fishing industry, fishing associations differed from fishing lodges in the degree of control that the state could exercise over them. The Fishing Association Law required fishing associations to register with the local and central governments immediately after their formation. All fishing organizations founded prior to the enactment of the Fishing Association Law would have to reorganize and register within six months. The law also required fishing associations to forward information on their leadership personnel, membership, finances, and other affairs to the government on an annual basis. The associations were obligated to make similar reports on fishery production and their handling of fishing disputes. According to the Fishing Association Law, the state had the power to inspect all associations' activities and financial affairs. Officials could impose orders and sanctions deemed necessary for the supervision of fishing associations and repeal any resolutions deemed illegal or harmful to the public interest.[108] With these provisions, the Chinese state could utilize fishing associations as an apparatus for managing China's fisheries and implementing plans for the industry's development.

The central government further asserted its power to regulate China's fisheries during the 1920s by codifying a formal legal definition of fishing rights. This legislation originated with the

Fishery Regulations (*Yuye tiaoli*) of 1926 and the Fishery Law
(*Yuye fa*), which went into effect alongside the Fishing Association Law in 1929.[109] Inspiration for the legislation came directly
from the national fishery law that Japan had promulgated in
1901.[110] According to the Nationalist government's Fishery Law
of 1929, anyone seeking access to fisheries in China's territorial
waters had to obtain permission and register with local officials,
who would forward this information to the provincial and central
governments. Only Chinese citizens were granted access to fishing rights in the nation's territorial waters. The legislation upheld
any fishing rights based on "contract or local custom" that existed
prior to its enactment. Yet the central government held final authority in determining the right to fish. The Fishery Law invested
government organs with the authority to prohibit any fishing activities when deemed necessary to protect the reproduction of
marine animals and plant life. Furthermore, all conflicts over fishing rights had to be handled through official adjudication. Objections to official settlements in these disputes had to pass through
the proper administrative appeals process.[111] On paper at least,
the Fishery Law made the central government the final arbiter in
decisions regarding the use of China's marine resources.

By exerting control over fishery management, the Fishery Law
and the Fishing Association Law made it possible for the Chinese
state to encourage more rational use of the marine environment
and thereby increase fishery production. With these laws, the
Chinese state claimed the power to transform institutions coordinating the use of fisheries whenever they were perceived as an
impediment to economic development. In principle, these pieces
of legislation superseded all rules and practices that the fishing
population used to regulate marine resources.

But in reality, the Chinese state's fishery legislation did not
radically alter the institutions that had existed in the Zhoushan
Archipelago's fisheries since the Qing period. Even after the Chinese Nationalist government consolidated its power in the Nanjing Decade (1927–37), it lacked the capacity to enforce provisions
in the Fishery Law requiring state regulation of access to fishing

grounds and official registration of fishing rights. Nor could the Chinese state make fishing associations into a mechanism for implementing its fishery development policies. On the contrary, the native-place lodges formed by fishermen's regional coalitions remained basically unchanged, and officials had no way to compel them to comply with the Fishing Association Law.

Indeed, most fishing lodges continued to function without regard for the legislation well into the 1930s. As late as 1934, the Jinghe and Nanding fishing lodges sent petitions to the Dinghai county government reporting the selection of headmen and directors in accordance with their old regulations.[112] The few native-place lodges that claimed to have reorganized as fishing associations did so as little more than a formality.[113] Even after changing their name, leadership of fishing associations stayed in the hands of former fishing lodge directors. The Taizhou Fishing Lodge's director Ge Liquan, for instance, handled the Linhai Fishing Association's affairs after it was founded in 1923.[114]

The leadership of Taizhou's Wenling Fishing Association exemplified the links between fishing lodges and fishing associations. The Wenling Fishing Association was founded in 1923, but the following year a dispute with the county government over collection of fishing registration fees led to its dissolution. The organization was not formally reinstated until 1929.[115] The Wenling Fishing Association's general director after its reorganization, Liu Boyu, had previously been police chief in various counties in Zhejiang and head of public security forces in Jiangsu's Songjiang county. During the early 1930s, Liu was director of the local-defense militia in Wenling county and manager of a fish brokerage.[116] A fish merchant named Chen Zhongxiu, who was head of Wenling's Songmen Chamber of Commerce, was also among the directors.[117] Actual control of the Wenling Fishing Association belonged to Bao Heng, who had greater knowledge of the fishing industry than Liu and Chen because he had directed native-place lodges for fishermen from Taizhou who migrated seasonally to Zhoushan.[118] Originally from Wenling, Bao Heng had graduated from the Ningbo Police Academy during the late Qing and attained renown in Taizhou after rescuing a group of Dutch sailors from a shipwreck. In recognition of this heroic deed, the Qing government awarded Bao a post as brigade vice-commander for the Green Standard military forces in Shipu. In addition to his honorary official credentials, Bao Heng engaged in the ice boat business, which gave him a considerable financial stake in the fishing industry. With the fall of the dynasty in 1911, Bao Heng left his military post and traveled to Ningbo, where he founded the Taizhou Lodge. During his long tenure as the lodge's director, Bao acted as a representative for Taizhou fishermen in their dealings with local officials.[119] Connections with fishing lodges were also apparent in the leadership of the Yin County Fishing Association, which came into existence in 1932. One of the association's directors was the Yongan Lodge's former director Shi Renhang. The association's supervisor, Xin Wenhuan, held leadership positions in several fishing lodges as well.[120]

Although state-imposed regulations did not significantly alter native-place organizations, fish lodge leaders gained considerable social capital through their official connections. Zhang Shenzhi, who served as director of the Yongfeng Lodge formed by ice boats from Yin county, obtained a provincial *juren* degree in the Qing civil-service examinations in 1901. Zhang was elected to the Zhejiang provincial assembly in 1907, and after the 1911 Revolution he handled the finances of Ningbo's provisional military government. Zhang later became a member of China's National Assembly, and in 1927 he briefly served as supervisor of the Zhejiang Maritime Customs. In between all these responsibilities, Zhang found time to oversee water-conservancy projects in the Ningbo region as head of the Yin County Water Conservancy Bureau (Yin xian shuili ju).[121]

In the late 1920s and early 1930s, official legislation did not significantly alter the collective organizations formed by native-place groups in the Zhoushan maritime region. All in all, fishing associations were essentially fishing lodges under a different name. For this reason, it comes as little surprise that they did not carry out measures to transform credit and marketing systems as spelled out in the Chinese state's fishery legislation. As was typical of many state efforts to regulate social organizations in Republican China, local elite leaders altered their titles to conform to officially prescribed nomenclature, but actual functions were highly resistant to state-mandated changes.

Conclusion

As the Zhoushan Archipelago's fishing industry continued to expand over the first three decades of the twentieth century, Chinese fishery experts allied with China's modern state endeavored to refashion interactions between society and the marine environment. Taking the reality of limited resources as their starting point, regionally based groups had crafted social institutions to protect profits derived from the marine environment by allocating access to fishing territories and resolving violent disputes. These arrangements confronted the problems of mounting competition and contests for control of scarce resources at the local level but did nothing to prevent overexploitation. In the 1910s and 1920s, the modern Chinese state's fishery experts put forth a different conception of the relationship between people and nature. Their developmental plans expressed unwavering faith that an alliance between science and the state would make exploitation of the environment optimally rational and efficient. Modern scientific and technical expertise, as they believed, could reconcile the goals of maximizing production and preventing overexploitation. These developmental assumptions extended from a conviction that centralized planning could manipulate nature for increased yields and unlimited utilization.

The ideas of Chinese fishery experts resembled the models of environmental management centered on efficiency and wise-use that dominated the United States and other Western nations during the late nineteenth and early twentieth centuries. Exponents of this approach held that scientifically trained experts employed by a central authority could apply their knowledge to formulate an objective, disinterested, and comprehensive view of environmental issues.[122] This form of environmental management, notes Kevin Frawley, entailed "a public policy framework designed and managed by experts who increasingly tried to apply scientific and economic principles to the efficient utilization of resources." The developmental objectives that informed this approach toward the natural environment in the early twentieth century sometimes demanded limiting use of natural resources, but only to ensure their availability for future exploitation.[123] In the realm of fisheries, as Arthur McEvoy points out, this understanding of relationships between nature and society led to the idea that "government and its scientific advisors could replace what civilization's carelessness had destroyed."[124] Fishery experts in Republican China shared this developmentalist bias, grounded in a belief that expert intervention could minimize waste and make the environment optimally productive. These similarities, of course, are not difficult to explain. Japan had adopted models of scientific fishery management from the United States and European countries during the Meiji period, and a generation later these environmental discourses found their way from Japan to China.

In twentieth-century China, as elsewhere, a drive to promote efficient utilization of resources through technology and scientific expertise was a central component of the modern project.[125] Yet despite the emphasis Chinese fishery experts placed on the impartial application of scientific knowledge in pursuit of the long-term public interest, they had to obtain sources of funding for their programs. Budgetary concerns entangled them in political maneuvering for tax revenues that all Chinese government agencies participated in during the early twentieth century. Hence, the need for funding shaped the direction of fishery policy in Republican China as much as a shared commitment to rational and productive exploitation of the environment. Without other reliable sources of financing, funding for fishery management agencies came primarily from taxes on fishing enterprises. In addition to making use of fish stocks more efficient, fishery development policies also generated more tax revenues and greater financial support for underfunded government agencies. On one hand, fishery management in Republican China promised to eliminate a major source of revenue fluctuation by making the ocean's resources easier to manipulate and harvest according to centralized plans, providing steady and uniform output. On the other hand, allocation and control of this income generated a great deal of contention and debate.[126]

All in all, the developmental initiatives advanced by Chinese fishery experts during the 1910s and 1920s yielded few tangible results. The Republican Chinese state simply did not have the funding or the administrative capacity to put the far-reaching reforms that they recommended into full effect. As a result, the question of whether state-led fishery development programs could actually increase extraction of finite resources while preventing depletion never had to be confronted. Despite their desire to transform patterns of interaction between nature and society, the plans of Chinese fishery experts ran up against major constraints. State legislation alone could not displace the local arrangements relied on by fishermen's native-place coalitions to coordinate use of Zhoushan's fisheries. With the plans still in the blueprint stage, fishery experts never had to question their faith that modern science and technology could overcome the effects of intensified human exploitation of the marine environment.

FOUR

Fishing Wars I: Sino-Japanese Disputes, 1924–1931

Marine fisheries do not conform to political boundaries. Many fish species migrate great distances, cutting across the borders of multiple nation-states. For this reason, international quarrels over fishing grounds are routine. During the 1920s and 1930s, the confluence of ecological changes in China's coastal waters and the broader marine environment of the East China Sea gave rise to disputes between China and Japan over Zhoushan's yellow croaker fishing grounds. Japanese trawlers paid no heed to regionally based fishing groups' proprietary claims to fishing territories. Moreover, Japanese vessels with modern harvesting technologies could easily outfish their unmechanized Chinese counterparts, and this seriously reduced Chinese catches. These Japanese incursions rendered useless the arrangements relied on by regionally based groups to coordinate the use of Zhoushan's fisheries.

In response to this Japanese pressure, fishing enterprises in Zhoushan called on the Chinese state to uphold their claims to fishing grounds and exclude Japanese vessels from these waters. These fishery disputes played out against the backdrop of tense relations between China and Japan in the late 1920s and early 1930s, as increasingly assertive Chinese nationalism ran up against

Japanese defense of its imperial privileges, especially in China. The limitations of prevailing tenets of international law and Japan's ability to apply diplomatic and military pressure in support of its fishing industry made it impossible for China to stop Japanese incursions into Zhoushan's waters. The depletion of already strained fish stocks accelerated as a result. What is more, transnational ecological disputes accentuated *internal* struggles for the control of marine resources. Provincial and central governments in China took advantage of Sino-Japanese fishery disputes to maneuver for a greater share of revenues collected from Zhoushan's fishing industry. As the array of Chinese state agencies that claimed the power to administer and tax Zhoushan's fisheries multiplied during the 1920s and early 1930s, the challenge of preventing violent competition for common-pool resources grew all the more difficult.

Modern Chinese Banks and Zhoushan's Fishing Industry

During the 1920s and 1930s, ongoing processes of economic integration with Shanghai and other urban centers put even greater demands on Zhoushan's marine environment. Modern Western-style Chinese banks with headquarters in Shanghai got involved in issuing loans to fish brokers in Zhoushan, supplementing financial networks that had emerged during the late Qing. The appearance of modern bank branches in Zhoushan paralleled the growth of financial institutions throughout the Chinese economy during the interwar period. The Commercial Bank of China (Zhongguo tongshang yinhang) first opened a branch office in Shenjiamen in 1922. Others quickly followed suit. By the early 1930s, the Bank of China (Zhongguo yinhang), the Communications Bank (Jiaotong yinhang), the Agricultural and Industrial Bank of China (Nonggong yinhang), the China Industrial Bank (Zhongguo shiye yinhang), and other modern banks had set up offices in, among other ports, Shenjiamen, Dinghai, Dongshajiao, and Shengshan to meet the need for credit during the fishing season.[1] The appearance of these modern financial institutions contributed to increased fishery output to satisfy the growing demand from urban markets.

The value of loans made by the Commercial Bank of China's Daishan office increased from 30,000–40,000 *yuan* to 60,000–70,000 *yuan* per month during the fishing season for large yellow croaker when the credit market was at its peak. Most of the loans went to fish brokers and fish-processing enterprises.[2] According to a 1935 report from the China Industrial Bank's Shenjiamen office, "Shenjiamen is a developing fishing port. Although it progresses with each year, all surplus capital is used [by fish brokers] to expand business, and thus ready capital is extremely scarce. Attracting deposits is quite difficult."[3] Bills of exchange issued by native banks and modern banks with branch offices in the Zhoushan islands enabled brokers to purchase the catch from fishing boats in the absence of ready cash.[4]

Fish brokers could usually obtain unsecured loans from native banks on the basis of their personal reputation.[5] Obtaining loans from modern banks, however, required collateral. Fish brokers had little property to offer as material securities except for the rice, hemp, and nets they supplied to fishing boats. Most modern bank branches in Shenjiamen let fish brokers use these materials as collateral but also required a guarantor to vouch for the borrower's creditworthiness. In the event that a fish broker defaulted on a loan, the guarantor shared responsibility for the bank's losses. Loan periods typically lasted from two to six months, with a monthly interest rate of approximately 0.8 to 1 percent. According to a report from the China Industrial Bank, most fish brokers were capable of making good on their debts, putting up nets as security when fishing activity reached a low point and redeeming them at the onset of the spring yellow croaker runs.[6]

During the 1920s and 1930s, the background of some of the most prominent fishing lodge directors reflected the growing influence of modern banks in Zhoushan's fishing industry. Liu Jiting, the manager of the Shenjiamen and Dinghai branches of the Bank of China, was director of the Shenjiamen Fish Brokers

Lodge (Shenjiamen yuzhan gongsuo) during the 1930s.[7] The founding director of the Commercial Bank of China's Dinghai branch, Zhang Xiaogeng, who invested in several other business enterprises related to the fishing industry, was a director of the Shenjiamen Fish Brokers Lodge as well.[8] One of the Yin County Fishing Association's directors, Cao Chaoyue, was also chairman of the business department in the Ningbo Industrial Bank's Shenjiamen Branch and a headman in the Yongan Lodge.[9]

The Ecological Dimensions of Fishery Expansion

In the Zhoushan region, economic integration experienced during the 1920s and 1930s appears to have facilitated substantial increases in fishery output. According to the best available estimate, total landings in the Zhoushan Archipelago amounted to approximately 60,000 tons in 1920; by 1936 the catch increased to 93,000 tons.[10] However, in the absence of reliable quantitative data, it is difficult to gauge the ecological impact of this expansion of fishery production on fish stocks.

In many fisheries under human exploitation, the largest fish are caught first. Continued fishing reduces the average size of fish because they do not have a chance to mature. Catching too many fish before they grow to maturity also impedes reproduction. These impacts can occur even in unmechanized fisheries like that of the Zhoushan Archipelago, which consisted of sail-powered junks well into the twentieth century. In Zhoushan during the 1920s and 1930s, fishermen used stow nets with extremely fine mesh, capturing large numbers of immature fish and eggs.[11] Fully developed large yellow croakers were considered a delicacy, but young and undersized fish were not eaten. Fishermen dried immature yellow croakers, known as "plums" (*meizi*) in Zhejiang, and sold them as fertilizer to farmers in Wenzhou and Fujian.[12] Fisheries thus helped meet the tremendous demand for fertilizer required to feed China's growing population, and soil nutrient loss contributed to fishery exploitation.[13] A Japanese investigation

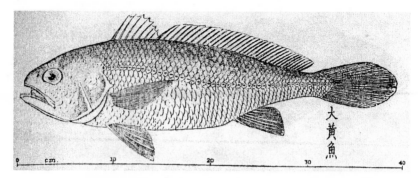

Fig. 7   Large yellow croaker. From: Read, *Common Food Fishes of Shanghai.*

from the 1920s noted, "The Zhoushan Archipelago's coastal waters are suited to the spawning of fish species; so all year there are
many immature fish under three or four inches."[14] Small, immature fish fetched a lower price on the market than large, mature
ones. Fishermen had to bring in bigger catches to recoup their
expenditures of labor and capital, which put greater pressure on
fish stocks.[15] Hence, this trend may well have contributed to
stock depletion by taking too many young fish before they could
grow to full maturity.[16]

During the 1910s and 1920s, moreover, fishing boats ventured
farther out to sea in search of new fishing grounds to maintain or
improve their catch-rates. As the Republican-period gazetteer of
Xiangshan county noted in 1927, "Those who cast their nets close
to shore catch few fish; so they must go to the far seas and deep
waters."[17] Similarly, according to a gazetteer of Taizhou prefecture published in the 1930s, "Fishers originally went to and from
the near seas, but now they have expanded to various places in
Zhenhai, Dinghai, Fenghua, and Xiangshan, while sturdy ones
even travel north to Chongming Island [at the mouth of the
Yangzi River in Jiangsu province] and south to Wenzhou and Fujian."[18]

Gradual expansion into more distant fishing grounds during
the Republican period is unmistakable, but the evidence does not
allow us to determine whether this trend resulted from human or
nonhuman causes. At least one contemporary attributed this

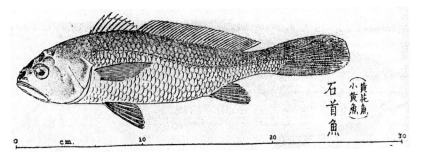

Fig. 8   Small yellow croaker. From: Read, *Common Food Fishes of Shanghai.*

movement into offshore fishing grounds to changes that resulted from the recent expansion of the region's fisheries. The 1920 gazetteer of Daishan Island explained:

> Those who have been involved in the fishing industry for a long time say that thirty or forty years ago large yellow croakers followed the tides and came to islands close to shore. At that time, there were not many boats, their fishing methods were crude, and catching fish was relatively easy. Now the number of fishing boats has increased, fishing gear is better, and the fish hide in the deep waters.[19]

The growth of fishing activity since the late nineteenth century, in this view, heightened competition for limited inshore fish stocks, forcing boats farther out to sea in search of higher yields and greater profits. Nevertheless, it is impossible to rule out the possibility that the abundance and distribution of fish stocks changed in response to "natural" environmental fluctuations, such as changes in weather conditions, water temperature, or ocean currents.[20] What is certain is that, to borrow W. Jeffrey Bolster's phrase, "the inability of the ecosystem to produce the volume desired by harvesters" led to a perceived shortage of fish, prompting fishing enterprises to actively seek out other stocks.[21] There is no way to be sure that human exploitation had depleted fishing grounds close to shore, or if naturally occurring changes in the ecosystem had made it impossible to satisfy a growing demand for fish. But for whatever the reason, fishermen in Zhoushan wanted more fish than they could catch in near-shore waters.

Fishery Expansion in the Shengsi Islands

By the 1910s, the main fishing grounds had shifted to waters off the Shengsi Islands, in the northernmost reaches of the Zhoushan Archipelago. Unlike the rest of the archipelago, which was part of Zhejiang's Dinghai county, the Shengsi Islands fell within the boundaries of Chongming county in Jiangsu. Fishermen from northern Zhejiang were following large yellow croaker to waters off the Shengsi Islands by the 1870s, after the productivity of fishing grounds near Xiaoyangshan Island declined.[22] But during the late Qing and the early years of the Republican period, only driftnet boats ventured into the distant waters in the Lüsi Sea off Jiangsu. A major turning point came in 1917, when fishermen discovered small yellow croaker and hairtail stocks in the Sheshan and Lüsi seas.

According to sources from the 1930s, the Jiangsu-Zhejiang Fishing Company's steamship *Fuhai* discovered abundant yellow croaker stocks while patrolling for pirates off Shengshan Island and told fishermen about them.[23] These previously unexploited resources attracted boats into waters over one hundred miles from the archipelago's outermost islands.[24] Lured by these new fishing grounds, ever-greater numbers of boats came to the Shengsi Islands over the course of the 1920s (see Table 4.1).[25]

Table 4.1
Fishing Boats in the Shengsi Islands

Island	1899	1934
Shengshan	80	268
Gouqi	90	354
Huaniaoshan	50	222
Lühua	30	108
Sijiao and Majishan	130	855
Xiaoyangshan	30	57
TOTAL	410	1,864

SOURCE: Cheng Tiyun, "Jiangsu waihai shan dao zhi."

It did not take long for the Shengsi Islands to turn into a booming fishing center. The early 1930s marked the heyday of fish brokerages in the Shengsi Islands, with the opening of more than ten of these businesses.[26] As in the rest of the Zhoushan region, most fish merchants in Shengsi were seasonal migrants. Of the eighteen fish brokerages on Shengshan Island in 1932, only seven had settled there permanently. The rest came from sojourning groups (*wai bang*) that returned to their native place at the end of fishing season.[27] Many of the fish brokerages that did business in the Shengsi Islands on a permanent basis had been set up by settlers from other islands, such as Daishan, closer to the Zhejiang mainland.[28]

Japanese Mechanized Fisheries in the East China Sea

On the other side of the East China Sea, Japan's mechanized fishing industry replicated the environmental trends seen in Zhoushan's fisheries, but on a far more massive scale. In the late nineteenth century, the Japanese government had successfully embarked on an ambitious modernization program to break into the ranks of the world's imperial powers. From the 1890s, Japan relied on this newly acquired industrial and military might to extend its hold over Taiwan and Korea, as well as parts of Manchuria and other areas of north China. Development of Japan's fishing industry went hand in hand with the country's industrialization at home and imperial expansion abroad.

The modernization of Japan's fisheries began during the late Meiji period with the introduction of steam-powered fishing boats from Western Europe. Steel-bodied otter trawlers were first introduced to Japan from England in 1908, and this technology spread rapidly in Japan's fishing industry. Otter trawlers were modern deep-sea fishing vessels that used large boards or metal plates called otter boards to spread the trawl nets. As steam-powered boats towed trawl nets along the sea bottom, water pressure on the otter boards' surface forced the boards outward and kept the net's opening spread taut.[29]

Initially, steam-powered otter trawlers fished in waters off the coast of Japan, leading to confrontations between traditional fishers and the operators of mechanized vessels. The Japanese government settled domestic conflicts over fishing grounds by forcing modern vessels to move into more distant seas. In 1911 the Japanese government placed restrictions on the use of otter trawlers in its coastal waters. With the enlargement of these restricted areas in 1912, the Japanese government limited the use of otter trawlers to waters west of 130 degrees longitude. Following the enactment of this legislation, Japanese otter trawlers fished primarily for sea bream (J. *tai*, Ch. *diaoyu*) in the East China and Yellow seas.

Seeking to prevent diminishing yields, in 1923 the Japanese government limited the number of licensed otter trawlers to seventy vessels, all of which had to exceed 200 tons and be capable of speeds over eleven knots. These limits did little to offset declining catches, however, and the Japanese fleet's concentration on prime fish species such as bream soon led to their depletion. Between 1925 and 1927 many Japanese vessels began fishing for yellow croakers (J. *guchi*), which Japanese considered lower-quality species and used mainly as raw material for processed fish products. In the early 1920s, Japan's mechanized trawlers headed for waters off the coast of China where yellow croakers came to spawn. Changes in the amount of bream and croaker caught by Japanese trawlers during the 1920s reflected this shift in target species (see Table 4.2). Japan's government actively encouraged this movement by granting subsidies to enterprises that engaged in pelagic fisheries and fishing in foreign waters.[30]

Motorized drag-net boats, which made up another sector of Japan's modern fishing fleet, followed a similar pattern. Unlike otter trawlers, which were imported from foreign countries, drag-net boats combined internal combustion engines with traditional Japanese fishing techniques. This fusion of foreign and domestic

Table 4.2
Number of Boats and Catches of Bream and Croaker by
Japanese Trawlers in the East China and Yellow Seas, 1921–28
(thousands of tons)

Year	Boats	Bream (*rengodai*)	Croaker (*guchi*)
1921	48	5.49	5.00
1922	—	4.94	7.06
1923	68	4.11	7.75
1924	—	2.47	8.59
1925	65	1.64	10.19
1926	67	1.58	14.62
1927	65	0.68	16.61
1928	64	0.17	21.32

SOURCE: Kibesaki, "Fundamental Studies on Structure and Effective Management of the Demersal Fish Resources in the East China and the Yellow Sea," 75.

technologies required far less capital than otter trawlers and gained wide acceptance in the Japan's fishing industry. Following the earliest experimentation with this technology in 1920, the number of drag-net boats in operation reached 200 in just two years. After another round of conflict with traditional fishermen in Japan's coastal waters, in 1921 the government limited drag-net boats to waters west of 130 degrees longitude.[31] Since, however, the Japanese government did not enact legislation to limit the number of drag-net boats, their numbers jumped to 300 boats in 1926 and 1,000 by 1937. Like steam trawlers, drag-net boats initially fished for bream and other high-priced species in waters close to Japan. Following the decline of these fish stocks, Japanese drag-net boats also began to fish mainly for yellow croakers.[32]

Over the course of the 1920s, catches of prime species such as sea bream by Japan's mechanized fishing fleet declined, and the previously less-exploited yellow croaker took their place. Between 1921 and 1929, croaker species jumped from 20 to 30 percent of the overall catch by Japanese steam trawlers in the East China and Yellow seas. In 1937 these species accounted for over 40 percent of the total catch. This trend reflected the major change in the Japanese fleet's target species following the depletion of prime fish stocks.[33] As a result, Japanese vessels moved ever closer to the yellow croaker fishing grounds off the outermost islands of the Zhoushan Archipelago.

Sino-Japanese Fishery Conflicts, 1925–27

Prior to the mid-1920s, the activities of the Japanese fishing fleet did not have a direct impact on Zhoushan's fishing industry. The 1924 edition of the Dinghai county gazetteer noted that Japan exported a particularly large amount of sturgeon (Ch. *qingzhan*) to China but pointed out that "in Dinghai sturgeon is considered a type of miscellaneous fish. No one specializes in catching it, and not much is produced. The reason for this is that sturgeon is found in deep waters and cannot be caught with traditional fishing methods."[34] As long as Japanese vessels stayed in distant waters and Chinese boats close to shore, the Japanese and Chinese fleets did not compete. Conflict between Chinese and Japanese fishermen over access to Zhoushan's fishing grounds emerged only after the Japanese fleet began to focus on catching yellow croaker during the mid-1920s. Japanese trawlers' shift to these new target species drew them to the waters off Zhejiang and Jiangsu where yellow croaker came to spawn during the spring and summer months. Chinese boats had ventured into these waters only recently, as pressures on fishing grounds closer to shore intensified during the early twentieth century.

The convergence of environmental influences from China and Japan soon gave rise to international contests for control of Zhoushan's yellow croaker fishing grounds. Chinese fishing interests in Zhoushan claimed that Japan's fishing fleet had violated China's sovereignty by fishing in its territorial waters. Their Japanese competitors argued otherwise. Consequently, Sino-Japanese fishing disputes in the 1920s and 1930s centered on contested definitions of territorial waters in international law.

Japanese boats first arrived in waters off the Shengsi Islands in the spring of 1925 during yellow croaker season. The unequal treaties signed between China and foreign powers during the nineteenth century had made Shanghai an open port for international trade. These imperialist privileges gave Japanese trawlers the ability to land and sell fish in Shanghai without impediment, which they did on a growing scale in the 1920s. The arrival of foreign competitors resulted in panic among the city's Chinese fish merchants.[35] Shanghai fish merchants' Dunhe Lodge gathered all its members for an emergency meeting with the Shanghai Chamber of Commerce (Shanghai zong shanghui) and Haizhou Fishery Training Institute director Wang Wentai.[36] This outcry came on the eve of massive antiforeign boycotts connected with the May 30 Movement, in which the Chamber of Commerce played a major role.[37] The same nationalist rhetoric that infused the boycott movement characterized the fishing dispute. Japanese boats, it was maintained, had "violated international law and encroached on [China's] sovereign rights over territorial waters."[38] The Shanghai Chamber of Commerce urged China's Ministry of Foreign Affairs to protest to the Japanese embassy and requested vessels from Jiangsu province's coast guard to patrol fishing grounds off the Zhoushan Archipelago to "protect our sea power and support the fishing industry."[39] The Shanghai business community's investments in fishing enterprises explain its desire for government protection against foreign competition. The Shanghai Chamber of Commerce's director, Fang Jiaobo, hailed from a prominent merchant lineage in Zhenhai county, whose substantial business portfolio included investments in fishing enterprises.[40] In addition to his position in the Chamber of Commerce, Fang Jiaobo was also head of the Dunhe Lodge.[41] From 1928 to 1936, fishing lodge director Zhang Shenzhi was also director of Shanghai's Ningbo Native Place Association (Ningbo lü Hu tongxianghui), with which Fang Jiaobo and many other Shanghai banking figures were closely connected.[42]

These business interests were influential enough for the Chinese government to take diplomatic action. Immediately after receiving word of the Japanese fleet's incursion off the Shengsi Islands, China's Ministry of Foreign Affairs protested to the

Japanese embassy, charging that Japan's fishing vessels had "fished within China's territorial waters and purposely violated China's fishing industry and maritime rights (*haiquan*)."[43] Japan's diplomatic representatives countered by insisting that Japanese boats fished only outside China's territorial waters in seas open to vessels from all countries.[44]

Territorial Waters in International Law

The Sino-Japanese disputes over Zhoushan's yellow croaker fishing grounds coincided with global debates over the status of territorial waters in international law. Most countries recognized coastal states' sovereignty over territorial seas extending beyond their immediate shores, but the *breadth* of these waters was open to question. Powers such as Great Britain had adhered to a three-mile territorial limit since the nineteenth century, but this principle was not unanimously accepted.[45] For this reason, the issue of territorial waters occupied a prominent place on the agenda of the Committee for the Progressive Codification of International Law formed by the League of Nations in 1924. Replies to the committee's Schedule of Points revealed major differences of opinion regarding this question. Only Japan and Great Britain, countries able to outfish their neighbors, voiced unconditional support for a maximum three-mile limit.[46]

The League of Nations' First International Conference for the Codification of International Law convened at The Hague from March 13 to April 12, 1930. The conference failed to adopt a convention on territorial waters, producing only a set of draft articles that it distributed to various countries to reach an agreement in the future.[47] Following this failure to produce a consensus, territorial waters remained a point of ambiguity in international law. All countries agreed that claims to territorial seas could extend up to three miles, but no rule existed for the maximum limit.[48] China and Japan took remarkably similar positions on the territorial waters issue. Japan followed the British Commonwealth in backing a three-mile limit, and China expressed approval for a three-mile limit "in principle."[49]

China's earliest pronouncement on the breadth of its territorial waters came in September 1921, when the Ministry of the Navy concluded that China should follow other countries by defining its territorial seas as three nautical miles.[50] During the mid-1920s, fishing disputes with Japan brought this issue back into official discussion. China's National Assembly passed a resolution in May 1925 calling for the demarcation of territorial waters "to consolidate national power, rectify the fishing industry, and defend against foreign aggression." The National Assembly did not extend China's territorial waters beyond three miles but ordered definition and enforcement of maritime boundaries to prevent Japanese encroachments on fishing grounds off China's coast.[51] These measures fell far short of recommendations from fishing organizations in Zhejiang, which urged the Chinese government to claim waters 100 miles from shore.[52]

Fishing Protection and Fiscal Revenues

As diplomats tried to define China's territorial waters in the international arena, fishing disputes with Japan gave China's domestic powerholders access to revenues extracted from Zhoushan's fisheries. During the 1920s, competing warlord military forces in Zhejiang and Jiangsu claimed the power to defend Chinese fishing boats against Japanese encroachments. This did not reflect concern for the fortunes of Chinese fishing enterprises. Rather, provincial military leaders wanted to "protect" fishing boats to collect fees for this service.

Sino-Japanese fishing controversies followed on the heels of the 1924 Zhejiang-Jiangsu War waged between military forces in control of the two provinces.[53] Struggles over the collection of fishing taxes were one of the many ways provincial strongmen squeezed funds from Chinese businesses to meet their military expenses.[54] Claiming the need to prevent further Japanese encroachments, Zhejiang naval forces dispatched gunboats to patrol fishing grounds off Jiangsu and set up a Fishery Defense Affairs Office (Baowei yuye shiwuchu) in Shanghai to levy taxes on fishing enterprises.[55] These initiatives directly threatened the income that fishing lodges used to finance the mutual-defense forces that secured their claims to profits from Zhoushan's fisheries. The Zhejiang navy's "fishery defense" initiative met with strong opposition from the Yongfeng and Yongan lodges, which represented ice boats and large pair boats from Yin county, respectively. The lodges objected that the fishery defense initiatives were simply an excuse to supplement Zhejiang's budgetary shortages with fees collected from fishing boats. Since the lodges maintained their own militia forces, they had no need for outside protection.[56]

Unwilling to forfeit a lucrative source of revenues to their Zhejiang enemies, regional military forces in Jiangsu also moved to take control of taxes from Zhoushan's fishing industry. Hoping to obtain badly needed funding, Chinese fishery experts employed at research institutes in Jiangsu collaborated in these efforts. In June 1924, Jiangsu authorities deputed the head of the Haizhou Fishery Training Institute, Wang Wentai, to set up Jiangsu's own Fishery Defense Bureau (Jiangsu baowei yuye ju).[57] The new agency brought in the fishery specialists Li Shixiang and Hou Chaohai as consultants.[58] Wang Wentai's proclamation announcing the opening of the Fishery Defense Bureau explained that the disorganized and fragmented character of Jiangsu's fisheries had allowed Japanese trawlers to enter waters off the Shengsi Islands unimpeded. Furthermore, "outside parties" used protecting fishermen against foreign encroachments as an excuse to levy fees on fishing boats, and these ploys were bound to have a negative influence on the fishing industry.[59]

According to Wang Wentai, the Jiangsu Fishery Defense Bureau intended to create a prosperous future for fishing enterprises and would not saddle them with additional financial burdens. Wang claimed that Jiangsu's levies differed from the ones demanded by Zhejiang's naval forces, which used fishery protection as a way to extort fees. Nevertheless, Jiangsu's budgetary constraints made it necessary to collect fees "obtained from fishermen and fish merchants and used for the purpose of defense and rejuvenation." The Fishery Defense Bureau planned to construct a modern fishing port on Shengshan Island to strengthen the Jiangsu fishing industry's ability to compete with foreign vessels. According to Wang, the bureau would also encourage fishermen to form officially supervised associations and enact measures to protect the reproduction of fish stocks. The Fishery Defense Bureau's branch offices on Shengshan Island and in other fishing ports would send patrol vessels to settle conflicts between fishing groups from northern and southern Jiangsu and guard against incursions by Japanese vessels.[60]

Wang Wentai pledged that the Fishery Defense Bureau's patrols would cooperate with the mutual-defense forces formed by fishing groups' collective organizations (*fatuan*). In actuality, the Fishery Defense Bureau challenged the collection of fees by the Jiangsu-Zhejiang Fishing Association, which had come under the control of fishing lodge leaders after Zhang Jian's fishing company folded in the early 1920s. After its founding, the Fishery Defense Bureau intended to take over the Fishing Association's task of safeguarding the silver used in commercial transactions between fishermen and fish merchants. Arguing that the Jiangsu-Zhejiang Fishing Association's steamships had been purchased with provincial tax revenues, Jiangsu province sent its coast guard units to reclaim the ships and hand them over to the Fishery Defense Bureau, which would use them to protect fishermen against Japanese threats.[61]

Military developments soon brought Zhejiang's and Jiangsu's struggles for revenues from Zhoushan's fisheries to an abrupt halt. After occupying Shanghai in 1926, the warlord Sun Chuanfang replaced all existing provincial fishery administration agencies with a new Jiangsu-Zhejiang Fishery Affairs Bureau (Jiang-Zhe yuye shiwu ju).[62] The fishery expert Li Shixiang was enlisted as a technical advisor. To take control of fishing taxes, Sun Chuanfang dissolved the Jiangsu-Zhejiang Fishing Association and claimed the organization's steamships for the Fishery Affairs Bureau.[63] Fishing lodge leaders objected to this attempt to seize the Fishing Association's revenues, which amounted to 60,000 *yuan* annually, but eventually gave in. In August 1926, fishing lodge leaders gathered at the Yongfeng Lodge's Shanghai offices and handed over the Fishing Association's account books to the Jiangsu-Zhejiang Fishery Affairs Bureau.[64] After taking over the association's steamships, the Fishery Affairs Bureau levied a tax of 8 *yuan* on every catty (*dan*) of fish sold, or 30 *yuan* from each fishing boat. The bureau also collected customs duties of 15 *yuan* from every ice boat that entered the ports of Ningbo and Shanghai.[65]

The Evolution of Nationalist Fishery Administration

By 1927, China's Nationalist regime had defeated its military rivals in the Lower Yangzi region. Like the earlier warlord regimes, the Nationalist government relied on taxation of trade in Shanghai and other economic centers as its major source of revenue. Once the Nationalists began to govern Zhejiang and Jiangsu, they pursued the goal of consolidating control over fishing taxes just as vigorously as had Sun Chuanfang. The Nationalists took over Sun's Jiangsu-Zhejiang Fishery Affairs Bureau, placing it under the Ministry of Finance and assigning its steamships the task of preventing foreign encroachments on China's territorial waters. The Fishery Affairs Bureau still collected customs duties from fishing boats and levied a 5 percent tax on the price of fish sold by brokers. As with the Nationalist regime's decision to leave collection of land taxes in the hands of provincial governments, centralizing control over fishing tax revenues came with concessions to Zhejiang and Jiangsu. The central government took 40 percent of the revenues brought in by the Fishery Affairs Bureau, and the two provinces split the remaining 60 percent.[66] Zhejiang also convinced the Nanjing government to alter provisions in the Fishery Affairs Bureau's regulations that gave it authority to issue permits to fishing boats and oversee fishing militia organizations, arguing that these responsibilities—as well as the fees that came with them—belonged to the provinces.[67] Other interested parties also tried to influence the Nationalist government's reorganization of the Fishery Affairs Bureau to further their own agendas. Fishing lodges urged the Nanjing regime to dissolve the Fishery Affairs Bureau, claiming that its levies placed an onerous burden on fishing enterprises.[68] Fishery experts' professional organization, the Republic of China Fishery Studies Association (Zhonghua minguo shuichan xuehui), pressed the central government to let the Ministry of Agriculture and Mining share jurisdiction over the Jiangsu-Zhejiang Fishery Affairs Bureau with the Ministry of Finance. The Fishery Studies Association had successfully persuaded the Nationalist government to form a central fishery administration office staffed by many of its members under the Ministry of Agriculture and Mining. Consequently, putting this ministry in charge of the Fishing Affairs Bureau would give fishery specialists a way to gather funds for their reform initiatives.[69] In the end, however, none of these proposals were accepted by the Nationalist regime.

Efforts to influence government policy in order to secure a larger share of the profits from the fishing industry also surfaced at the Jiangsu-Zhejiang Fishery Reconstruction Conference, held in August 1928 to chart the future of fishery administration.[70] Industry representatives and fishery experts took advantage of this conference to lobby for measures to strengthen China's fishing industry against Japanese competition.[71] The Yongfeng Lodge's director Zhang Shenzhi asked for a reduction in domestic fishing taxes and higher duties on foreign imports to protect Chinese producers. Fishery specialists advocated a state-run marketing facility in Shanghai and a fishery bank to underwrite cooperatives. In addition, fishery experts and industry representatives urged the central government and the provinces of Zhejiang and Jiangsu to form a fishing harbor planning commission (*yugang sheji weiyuanhui*) to construct a modernized fishing port on Shengshan Island and make the blueprint for fishery development envisioned in Sun Yat-sen's *Industrial Plan* a reality.[72] Faced with Japanese

incursions on offshore fishing grounds, fishery experts and industry representatives called on the Nationalist government to increase domestic production by promoting more efficient exploitation of marine resources.

Sino-Japanese Fishing Disputes in the Nanjing Decade, 1927–37

In the spring of 1930, the Nationalist government received a new
flurry of petitions from fishing industry representatives in Zhejiang and Shanghai decrying the actions of Japanese trawlers,
which continued to catch fish in waters off the Shengsi Islands
and bring them to Shanghai for sale. According to these appeals,
Japanese vessels had violated international law and fished in
China's territorial waters. Chinese fishing interests called on the
Nationalist government to negotiate with the Japanese consulate
to protect the livelihood of Zhejiang's fishermen and bring an end
to this "national disgrace."[73] Zhuang Songfu, who was now vice director of the Nationalist government's Huai River Conservancy
Commission, cabled his fellow Fenghua compatriot Chiang Kai-
shek and requested central-government intervention to protect
the fishing industry,

I, Songfu, recently traveled home to find the Ningbo fish market on the
Yong River in a great panic because Japanese trawlers have encroached
on Jiangsu's and Zhejiang's Sheshan, Haijiao, and Langgang seas. These
huge vessels have large nets and catch all the fish indiscriminately, and
so from winter to spring the income of our country's fishers has greatly
declined.[74]

With this appeal, Zhuang drew on the native-place ties that had
mobilized Fenghua fishermen for the 1911 Revolution to garner
assistance for fishing enterprises.

Protests against Japanese encroachments came just as the Nationalist government was engaged in a diplomatic drive to reassert China's sovereignty and eliminate foreign privileges imposed by the nineteenth-century unequal treaties, such as
extraterritoriality and control over tariff rates. To the Nanjing

regime, Japan's alleged incursions into China's territorial waters were yet another violation of its national sovereignty. For this reason, the Nationalist government took Chinese fishing interests' appeals quite seriously. When the Ministry of Foreign Affairs protested to Japan's diplomatic representatives, the Japanese firmly held to their position that their trawlers fished only on the high seas and did no harm to Chinese boats.[75] In response, China's navy and the Ministry of Agriculture and Mining sent personnel to demarcate the nation's three-mile maritime belt. In this way, the Chinese government believed it would be able to refute the Japanese consulate's "excuse" that its fleet fished only on the high seas.[76]

Once again, the Nationalist government's measures for dealing with Japanese competition did not satisfy Chinese fishing enterprises. Particularly vigorous protests came from the Shanghai Municipal Chamber of Commerce (Shanghai shi shanghui), whose director Kui Yanfang owned eight fish brokerages in Shanghai.[77] The Chamber of Commerce reminded the Nanjing authorities that the issue of territorial waters had given rise to disagreements at the Hague Conference that remained unresolved. The Ministry of the Navy, which was responsible for demarcating China's territorial waters, believed that in light of China's geography the three-mile rule sufficed to define the limits of its maritime sovereignty. But, as the Chamber of Commerce stressed, adherence to the three-mile principle failed to prevent Japanese encroachments on Chinese fishing territories. For this reason, the Chamber of Commerce urged the navy to give the matter further thought and consideration.[78] By making these appeals, Chinese fishing enterprises urged the central government to take advantage of ambiguities in international law to protect their claims to offshore fishing grounds.

The Shanghai Municipal Chamber of Commerce's argument evoked a sympathetic response from Kong Xiangxi (H. H. Kung), the head of the Nationalist government's Ministry of Industry and Commerce. Fishery experts in the ministry's Department of Fisheries and Animal Husbandry (Yumu si) also encouraged an assertive stance against the Japanese fishing industry.[79] In April 1931, Kong Xiangxi attended the Shanghai Fishery Reform Promotional Assembly (Shanghai yuye gaijin xuanchuanhui), where he gave a speech detailing his proposals for China's fishing industry. All of Kong's ideas had long been advocated by Chinese fishery experts: tax relief for domestic fishing enterprises, 250,000 *yuan* in subsidies to encourage offshore fishing, officially organized fishery protection arrangements, 500,000 *yuan* for a central fishery experiment station, and government intervention to shield Chinese fishermen against encroachments by Japanese vessels.[80]

Responding to pressure from fishing enterprises and their political allies, in early 1931 China's National Assembly advocated extending the country's territorial waters to twelve miles.[81] The Ministry of the Navy warned that Nationalist China's diplomatic situation called for a more conservative position: "International conventions have set territorial waters as three miles. This country has just entered international organizations and certainly cannot increase them beyond three miles."[82] Although the Ministry of Finance wanted to set China's anti-smuggling boundaries at twelve miles (an accepted international practice), the navy held that this was "actually a type of administrative power outside the scope of territorial waters." In late April, following deliberations between the navy, the Ministry of Foreign Affairs, and the Ministry of the Interior, the Nationalist government set China's territorial waters at three miles with a contiguous anti-smuggling zone of twelve miles.[83]

Yet it soon became clear that simply defining the nation's territorial waters would not suffice to limit the Japanese fleet's access to fishing grounds off the Zhoushan Archipelago. Acting at Kong Xiangxi's request, the Ministry of Foreign Affairs demanded that the Japanese consulate forbid all Japanese vessels from overstepping China's maritime boundaries. Japan's chargé d'affaires responded by pointing out that fishing grounds no longer existed in such close proximity to the mouth of the Yangzi River. Both Chinese and Japanese boats fished between 50 and

100 miles from shore. For this reason, there were never instances of Japanese fishing in China's territorial waters off the coasts of Zhejiang and Jiangsu.[84] Indeed, subsequent Chinese investigations revealed that the gear used by Japan's mechanized fishing fleet could not be used in waters less than fifteen fathoms deep, making it impossible for them to fish within three miles of shore.[85]

The fact of the matter was that, as returns from inshore fish stocks declined, Chinese boats had set their sights on waters far beyond the three-mile limit in search of more productive fishing grounds. When Japanese vessels started to compete with them for yellow croaker stocks off China's coast, domestic fishing enterprises garnered government support by claiming that the Japanese had entered China's territorial waters and violated its national sovereignty. In actuality, not even the most liberal interpretation of the international law of the seas justified expanding China's territorial waters to include these disputed offshore fishing grounds.

Recognizing this difficulty, Kong Xiangxi turned to an indirect strategy to block Japanese competition for the Zhoushan Archipelago's marine resources. In February 1931 Kong brought a resolution before the National Assembly to "protect sovereignty and support the fishing industry" by preventing Japanese fishing activities in waters off Zhejiang and Jiangsu. The resolution clearly acknowledged that, according to international law, Japanese boats had not encroached on China's territorial waters:

Fishing boats catch fish far outside territorial seas. If one wants to prevent foreign boats' encroachments, then simply demarcating territorial waters will not solve the problem. Theories of the limits of territorial seas vary. Generally, all of them take three nautical miles measured from the low-water mark of a nation's outermost islands as their basis. In this near-shore range there are extremely few fish, and many of our country's old-style fishing boats fish in waters from forty or fifty to a hundred miles offshore. Boats from any country can fish in these places.[86]

Instead of redefining China's territorial seas, Kong proposed an alternative plan to limit the Japanese fleet's ability to compete with Chinese enterprises. First, the Ministry of Foreign Affairs would inform the Japanese consulate that since China and Japan had not signed a formal fishing treaty, Japanese fishing boats could not enter Chinese ports. Second, the Ministry of Finance would order maritime customs officials to prohibit any vessels carrying fish products from entering Chinese ports except for "proper merchant ships" that paid the required duties. Finally, Kong pointed out that the Ministry of Finance had already enacted regulations that banned boats under 100 tons from passing between Chinese and foreign ports to prevent smuggling. Since all the Japanese boats that landed fish in Shanghai fell below the minimum tonnage, customs officials could use this regulation as grounds for ordering them to leave port.[87]

On February 26, 1931, the Fourteenth National Assembly approved Kong Xiangxi's resolution.[88] To further protect Chinese producers against competition from foreign imports, on March 28 the Nationalist government waived all taxes on fishing enterprises. Once this policy went into effect, China's customs would charge a duty of 4.5 *yuan* on every catty of fish brought to port by Japanese boats, as stipulated in the two countries' tariff agreement, whereas Chinese fish products were not subject to any such levies.[89] Given the Nationalist government's thirst for revenues, this tax break came as a tremendous victory for the fishing industry.

Domestically, the elimination of fishing taxes deprived the Jiangsu-Zhejiang Fishery Bureau of its main source of funding. Responding to this change, the fishing lodges that previously ran the Jiangsu-Zhejiang Fishing Association tried to reassert their power to collect fees from fishing boats. Lodge leaders asked the Nationalist government for permission to take back the Fishing Affairs Bureau's steamships in order to organize mutual-defense convoys. The lodges pointed out that the Fishery Affairs Bureau no longer had any funding for these patrol ships, but Japanese encroachments and the threat of piracy made their protection more vital than ever. The central government needed to permit fishing lodges to gather funds to carry out defensive activities. The central government declined this request, deciding to place the

Jiangsu-Zhejiang Fishery Affairs Bureau and its steamships under the control of the Ministry of Industry, which had taken the place of the Ministry of Industry and Commerce. From this point on, the Jiangsu-Zhejiang Fishery Affairs Bureau was to be called the Jiangsu-Zhejiang Fishery Management Bureau (Jiang-Zhe yuye guanli ju).[90]

Internationally, restrictions on foreign fishing boats gave rise to protests from Japan's diplomatic representatives, who upheld the Japanese fishing fleet's rights to fishing grounds off the coast of China. Although the Chinese government claimed that it had enacted these customs measures to prevent smuggling, the restrictions did not apply to Chinese fishing boats that also engaged in these illicit activities. Nor did the restrictions deal with boats that sailed between ports in China's Pearl River Delta and the notorious smuggling centers of Hong Kong and Macao. Given these omissions, as the Japanese pointed out, the restrictions could not achieve their alleged goal of preventing smuggling.

In addition to questioning the validity of the Nationalist government's customs restrictions, Japanese consular representatives challenged the proposition that Japan's fishing fleet posed a genuine threat to Chinese fishing enterprises, "Motorized drag-net boats engaged in this type of fishing have never come under suspicion of smuggling. Moreover, they are working together with Chinese fishermen to open up the practically inexhaustible fishing grounds near China to supply the Chinese market. They have no desire to monopolize these fishing grounds." With the supposedly unlimited fishery resources off the coast of China and the "insatiable, constantly expanding demand" for fish products on the Chinese market, restrictions would not only bring losses to the Japanese but also hurt Chinese fish merchants and consumers. Since China did not have the capacity to fully exploit these "inexhaustible" resources, Japanese vessels had every right to exploit them as well.

Moreover, as the Japanese consulate reminded the Ministry of Foreign Affairs, China's customs policies violated treaty provisions stipulating that foreign boats could freely transport goods to and from specified Chinese ports. Hence, policies that prevented foreign boats from bringing fish caught on the high seas to treaty ports such as Shanghai illegally restricted trade.[91] Particularly strong protests against the Nationalist government's policies came from the Japanese-controlled ports of Qingdao and Andong in Manchuria's Guandong Leased Territory, where these restrictions threatened Japanese fishing enterprises.[92]

The Japanese consulate's objections also sparked controversy at the highest levels of the Nationalist government. Minister of Finance Song Ziwen (T. V. Soong) questioned the appropriateness of relying on customs restrictions intended to prevent smuggling to keep Japanese boats from landing fish caught on the high seas in Chinese ports.[93] After protracted negotiations over customs rates, Japan had agreed to grant China tariff autonomy on May 30, 1931.[94] From the Ministry of Finance's perspective, Kong Xiangxi's handling of the fishing dispute with Japan threatened this tenuous trade agreement. Mounting friction between China and Japan over Manchuria was an additional reason for the Ministry of Finance's misgivings.[95] Officers in Japan's Guandong Army had orchestrated the assassination of the Manchurian warlord Zhang Zuolin in 1928 and displayed hostility toward Chinese movements to recover Japanese rights in the region during the early 1930s.[96] As Sino-Japanese relations grew increasingly tense, Maritime Customs personnel in Andong and other northeastern ports made it clear that they could not enforce the restrictions proposed by Kong Xiangxi if the Japanese resorted to armed resistance to land their fish products.[97]

Differences between Kong Xiangxi's Ministry of Industry and Song Ziwen's Ministry of Finance over the handling of fishing disputes with Japan derived from their contrasting bureaucratic priorities. In contrast to Kong Xiangxi's desire to lend government support to domestic industries, Song Ziwen favored obtaining greater tax revenues to pay off the Nationalist government's debts and meet its military expenditures.[98] Since the Nationalist government had waived fishing taxes and placed the Jiangsu-Zhejiang Fishery Management Bureau under the Ministry of

Industry, officials in the Ministry of Finance had little to gain by enforcing the controversial restrictions on Japanese fishing boats. From the Ministry of Finance's perspective, restrictions on Japanese fish imports would only decrease the revenues it gained from customs duties. For this reason, the Ministry of Finance urged holding off on implementing these customs measures and finding a diplomatic solution to the fishing dispute with Japan. The Ministry of Foreign Affairs, which had to deal with the Japanese consulate's protests, supported this cautious position.[99] In May 1931 the Ministry of Industry agreed to delay the implementation of restrictions on foreign fishing boats.[100]

With the customs restrictions on hold, the Ministry of Industry turned to yet another means of keeping Japanese fishing boats out of Chinese ports. The Ministry of Industry asked the Ministry of Finance to have the Maritime Customs detain all foreign fishing vessels that had not registered and obtained a permit from the Chinese government as required by the Fishery Law. The Ministry of Finance refused to carry out these measures, maintaining that China could not force foreign boats to register according to domestic legislation.[101] Eventually, Kong Xiangxi conceded that, as long as Japanese vessels passed through customs, they could enter Chinese ports.[102] The Ministry of Finance agreed to carry out customs restrictions on Japanese fishing vessels except in ports such as Andong, where "special circumstances" applied.[103]

Following this compromise, in late August 1931 the Nationalist government finally implemented its customs restrictions. Once in place, the regulations prevented Japanese fishing boats from entering the port of Shanghai.[104] But they could not stop Japanese boats from fishing for yellow croaker and other species in waters off the Zhoushan islands. Many Japanese vessels landed fish caught off the coast of Zhejiang and Jiangsu in the Japanese-controlled ports of Dalian and Qingdao, where Chinese merchants transshipped them in refrigerated containers to Shanghai for sale. Other Japanese vessels simply resorted to flying Chinese flags. Japan's fishing fleet found it even easier to get around these obstacles after the Shanghai Incident in early 1932, during which the Japanese military retaliated for violence against Japanese civilians in China after the Manchurian Incident of September 1931.[105] During the battles in Shanghai, Japanese fishing boats seized the opportunity to land their catch in the port without any impediment from Chinese customs officials.[106] As a report from the Jiangsu provincial government put it in 1934, "Japanese fishing steamers have vanished from Shanghai, but their fishing in our territorial waters and secret dumping have not ceased for a day."[107] In the end, Japan's position of diplomatic and military advantage made it impossible for the Chinese government to limit the Japanese fishing fleet's exploitation of Zhoushan's fishing grounds.

Environmental Consequences

By the mid-1930s, the combined demands of Chinese and Japanese fishing activities had led to a clear decline in the productivity of Zhoushan's yellow croaker fisheries. Although Japanese trawlers accounted for the bulk of these pressures, Chinese fishers contributed as well. The Huaniaoshan and Sheshan fishing grounds, which had first attracted attention in the early 1920s, already showed signs of declining catches as early as 1934.[108] An investigation carried out in 1936 found that the size of yellow croakers and other fish taken by Chinese stow nets in Zhoushan's fishing grounds fell well below the desirable market size.[109] The handful of Chinese-owned trawling enterprises that had opened in Shanghai to compete with Japan during the 1930s only contributed to these ecological pressures, since they fished in the same waters as traditional Chinese fishing boats and mechanized Japanese vessels.[110] Faced with persistent Japanese competition, Chinese fishing enterprises took as many fish as possible before they were gone, without any regard for the resource's preservation.

One Chinese fishery expert observed that in the East China Sea's yellow croaker fisheries during the 1930s, "the large-sized fish of the past have already decreased, and what were originally considered medium- and small-sized fish are now already seen as large-sized fish. Decreased size is a precursor to decreases in quantity, and therefore the future is certainly not promising."[111] As returns from the East China Sea's yellow croaker fishing grounds fell, Japanese fishing boats moved on to exploit the fisheries of the South China Sea starting around 1935.[112] Chinese fishermen in Zhejiang and Jiangsu, who lacked the mechanized technology necessary to move to these distant waters, persisted in their efforts to get more from a declining resource.

However, some contemporaries explained these changes in different terms. During the early 1930s, the Chinese fishery expert Yao Yongping attributed declining catches in fishing grounds near Daishan Island to the unpredictability of the marine environment and the fishing population's ignorance.[113]

Schools of fish follow the currents, and currents change along with ocean conditions. In the seas around Daishan Island the rise and fall of tides is great, and the currents have already shifted; so it goes without saying that the course of the schools of fish has also changed. However, our country's fishermen have not received an education and do not know about these principles. Once a fishing ground is discovered, it is not changed for centuries, and so it is no wonder that catches are declining and those engaged in fishing there are decreasing.[114]

Yao's statement clearly overlooked Chinese fishermen's steady movement into waters further from shore since the late nineteenth century. Instead, Yao claimed that even after variations in ocean currents altered the location of fish stocks, fishermen still concentrated on the same fishing grounds every year. The alleged ignorance and conservatism of the fishing population justified expert intervention. In Yao's perception at least, overfishing would no longer present a problem if expert researchers helped fishermen discover more productive fishing grounds. To sustain output, the state needed to establish fishery experiment stations to investigate ocean conditions, sending steamships or aircraft to find fish stocks and report their location to fishermen.[115]

Yet by stressing technological solutions to shortages of fish, this approach ignored the intensification of local conflict that had resulted from changes in the marine environment. Driven by persistent competition for declining fish stocks, fishermen from different native-place groups turned to violence to keep scarce resources out of the hands of others. A 1934 newspaper article about Daishan Island's fisheries noted, "In recent years, disputes during fishing season have been hard to avoid. It is just a matter of whether they are large or small." Military, police, coast guard, and militia forces carried out patrols day and night during fishing season to prevent outbreaks of violence and maintain order.[116] In October 1935 longstanding animosities between the Fenghua group and Xiangshan's Dongmen group flared during a brawl at a temple opera performance. Fishermen took up arms in preparation for a major confrontation, causing panic among local residents. Police forces in Daishan intervened in time to put down the conflict, forcing fishermen to return to their boats. The police put a stop to the opera and instructed shops that had closed due to fears of a major feud to reopen. Local governments still looked to the elite leaders of fishermen's native-place coalitions to handle such local conflicts. Officials assembled each fishing group's lodge directors at the police station in Daishan and called on them to prevent further disturbances.[117]

In other instances, however, fishing lodges had difficulty simply resolving conflicts that emerged within their own native-place groups. In July 1936 the Yihe Lodge failed to settle a feud among different groups of fishermen from Fenghua. Local police forces had to impose martial law (*jieyan*) to end the dispute, which resulted in one death and numerous injuries.[118] The regulations that native-place coalitions relied on to ensure peaceful exploitation of Zhoushan's fisheries had once kept these kinds of local conflict in check. By the 1930s, competition for a declining pool of resources placed these institutional arrangements in serious jeopardy.

Conclusion

Beginning in the mid-1920s, the ongoing expansion of Zhoushan's fisheries and the Japanese fishing fleet's overexploitation of fish stocks in the East China and Yellow seas brought them into competition for offshore fishing grounds. During the early twentieth century, Chinese and Japanese fishing enterprises coped with the exhaustion of fish stocks by seeking out and moving into new, untapped ecological frontiers. The crucial difference was that economic development, technology, and imperialist expansion gave Japanese producers the capacity to wreak ecological havoc on a far broader scale than their Chinese counterparts.

Of course, conflicts between technologically advanced vessels and unmechanized boats that move into more distant fishing grounds due to the decline of inshore stocks can occur even if the competitors come from the same country. This phenomenon explains the disputes between traditional fishers and mechanized trawlers in Japan's coastal waters during the early twentieth century. The Japanese government dealt with these domestic conflicts by enacting legislation that made sure the ecological burden of modern technology was shifted elsewhere. Following its exhaustion of highly valued fish species in the East China Sea during the 1920s, Japan's mechanized fleet set its sights on Zhoushan's yellow croaker stocks. At the same time, Chinese fishing enterprises took advantage of bigger boats and better gear to venture into waters well beyond China's three-mile territorial limit. Sino-Japanese fishing disputes reflected the Chinese fishing industry's increased reliance on offshore fish grounds as much as it did Japanese encroachments.

Given the tensions that characterized Sino-Japanese relations during the late 1920s and early 1930s, it comes as little surprise that this fishery controversy was settled through confrontation rather than compromise. During the early 1930s, China's Nationalist government viewed the question of fishing rights as one aspect of its larger drive to assert China's national sovereignty in the face of foreign threats. However, the Nanjing regime was not willing to risk Japanese reprisals by breaking with prevailing definitions of territorial waters in international law to achieve these goals. Kong Xiangxi showed a great deal of ingenuity in his efforts to circumvent these international limits by relying on customs policies to prevent the Japanese fleet from landing fish in Chinese ports. Yet the Japanese government's ability to support its fishing enterprises through diplomatic pressure and the threat of military force stifled all Chinese attempts to exclude foreign trawlers from waters off the Zhoushan islands. The unequal power relations that prevailed between China and Japan during the Republican period made it impossible to exclude the Japanese fleet from the offshore waters frequented by Chinese fishing boats. Under these circumstances, Chinese fishermen raced to catch as many fish as possible to keep them from ending up in Japanese nets.

Nevertheless, attributing the decline of Zhoushan's yellow croaker stocks to Japanese incursions tells only part of the story. The Chinese government did not seek to exclude Japanese vessels from its territorial waters out of a concern for the fish. Rather, it sought to limit foreign access to coastal fishing grounds to ensure their availability for the future development of China's fishing industry. Even as the Japanese fleet placed greater demands on fish stocks, the Chinese state's fishery agencies and their allies advocated measures to increase domestic fishery production through more efficient exploitation of the marine environment. Few of these calls for reform were ever put into practice. But if these developmental plans could be implemented, Chinese state actors—including fishery specialists starved for budgetary support—stood to gain expanded sources of funding. The drive to consolidate control over taxes on Zhoushan's fisheries stood as just one example of the modern Chinese state's effort to extract revenues at all levels of society with little regard for environmental consequences.

During the 1920s, domestic contests for control of Zhoushan's fisheries not only took place between native-place coalitions trying to maintain exclusive claims to specific ecological niches but also involved competing bureaucratic organs seeking to derive greater revenues from exploitation of the marine environment. As the following chapter demonstrates, the intervention of these overlapping state agencies in the management of Zhoushan's fisheries intensified conflicts between opposing native-place coalitions that flared during the 1930s.

FIVE

Fishing Wars II: The Cuttlefish Feud, 1932–1934

As Japan's mechanized fishing fleet intensified pressures on yellow croaker stocks, domestic trends further destabilized the social institutions that had once minimized violent competition for common-pool marine resources in the Zhoushan Archipelago's fisheries. Since the late Qing, regional coalitions of fishermen had staked proprietary claims to fishing territories and enacted rules coordinating their use. When contests for the control of scarce resources gave way to conflict, the leaders of native-place lodges stepped in to mediate, keep the peace, and prevent the loss of profits to violent disputes. During the 1920s and early 1930s, however, long-term environmental transformations combined with changing socioeconomic conditions to destabilize these arrangements. Between 1932 and 1934, protracted feuding raged between fishermen from different regions of Zhejiang who harvested cuttlefish in Zhoushan's waters, and local forms of regulation proved totally incapable of settling these disputes. As violence escalated, fishery experts from China's central and provincial governments intervened to impose a rational, scientific solution to the feud. Yet their involvement only intensified the level of conflict, as fishery management gave way to bureaucratic struggles for revenues.

The Ecology of Cuttlefish

Cuttlefish (*Sepiella maindroni*, Ch. *moyu* or *wuzei*) make their home in relatively deep waters during the winter but come to inshore waters near the Zhoushan Archipelago to lay their eggs and spawn during spring and early summer.[1] Since the fishing grounds where cuttlefish congregated were too shallow for Japan's mechanized trawlers, only sail-powered Chinese boats fished for this species. Each spring, cuttlefish runs lured fishermen from the Ningbo region north to fishing grounds around the islands. During the 1880s, an American consul in Ningbo stated: "One of the principal, and, perhaps most profitable industries of this district is the ming fu, or cuttle-fish trade."[2] According to his estimate, 14,000 boats took part in cuttlefish season. Of these, 12,000 came from Ningbo and its vicinity, and another 2,000 came from "small hamlets along the coast" and the Zhoushan Islands. The approximately 84,000 fishermen were "drawn, principally, from the agricultural classes, receiving each from $8 to $20 for their services during the season."[3] In the early twentieth century, another foreign observer recalled, "thousands of fishing boats" from Shenjiamen and Ningbo "invaded" the Shengsi Islands during the cuttlefish season from April to June. "When this short season is over, the fishermen—and the flies—disappear and their few inhabitants are left to themselves."[4]

As in the other sectors of Zhoushan's fishing industry, the migrant groups harvesting cuttlefish stocks divided resource space according to native-place affiliation and specific technologies. Trawl-net (*tuowang*) fishermen, most of whom came from Jiangshan village in Yin county, claimed the waters around Gouqi, Shengshan, Huaniaoshan, and Lühua islands. Permanent residents of the Shengsi Islands fished with torch nets (*zhaowang*), which took advantage of cuttlefishes' affinity for light, in places where they could attach their gear to rocks jutting out into the sea. Seasonal migrants from Zhenhai county and fishermen from the Ningbo area who had settled on Huanglong and Sijiao islands

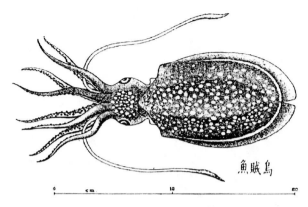

魚賊烏

Fig. 9   Cuttlefish. From: Read, *Common Food Fishes of Shanghai.*

fished with stow nets in spots where the ocean floor was not too rocky. Each of these native-place groups used technologies suited to different environmental niches, making it easier for them to divvy up access to resource space. When competition for cuttlefish escalated into disputes, native-place lodges carried out mediation, thereby minimizing losses that could result from these violent confrontations.[5]

Jiangsu Fishery Management Programs in the Shengsi Islands

By the early 1930s, regionally based lodges existed alongside the fishery management agencies that provincial governments had set up in the Zhoushan Archipelago. The most influential of these fishery offices emerged in response to mounting Japanese competition for offshore fishing grounds, which prompted Jiangsu to implement policies to reform and develop the province's fishing industry. These plans for strengthening Jiangsu's fisheries reflected the longstanding developmental objectives of Chinese fishery experts.[6] Jiangsu province's fishery reform program called for expanding the available resource base by searching for new fishing grounds, promoting the use of new types of gear, offering incentives to boats that fished on the high seas, and initiating

Fig. 10   Drying cuttlefish. Chinese National Relief and Rehabilitation Administration
Photographs, Box 4, no. 1818, Hoover Institution Archives.

research on aquaculture. Jiangsu's fishery management agencies
also called for measures to prevent wasteful, inefficient use of marine resources by limiting the use of stow nets with excessively
fine mesh and setting standards for other types of gear.[7]

As part of its fishery reform initiatives, in 1930 the province
founded the Jiangsu Provincial Fishery Experiment Station (Jiangsu shengli yuye shiyanchang) on Shengshan Island. Wang Wentai
was appointed as the director, and Zhang Zhuzun (another Japan-
trained fishery expert) was put in charge of investigating ocean
conditions.[8] The Jiangsu Provincial Fishery Experiment Station
was affiliated with the Fishery Direction Office (Yuye zhidaosuo),
which was set up to implement reforms at the local level. The Di-

rection Office opened on Shengshan Island in the summer of 1931, with the fishery expert Wang Liansheng as its director.[9] Its first task was to undertake an investigation of the fishing population in the Shengsi Islands, relying on divisions of Jiangsu's coast guard to assist with inspections for compiling household registers.[10] By registering the Shengsi Islands' population, the Fishery Direction Office would enable Jiangsu to secure additional tax revenues from fishing enterprises.

The Fishery Direction Office worked with Jiangsu's Chongming county, the administrative unit in which the Shengsi Islands were located, to set up state-sanctioned fishing organizations in line with the Nationalist regime's Fishing Association Law. In January 1932, Wang Liansheng joined with the chairman of Chongming's Nationalist Party branch and the head of the Chongming County Fishing Association to found the Chongming County Fishing Association's Shengshan Branch Association (Chongming xian yuhui Shengshan fenhui) on Shengshan Island.[11] These fishery management initiatives in the Shengsi Islands, like so many of the Chinese state's efforts to control local society during the Republican period, depended on the support of influential local elites. Yang Youcai, a native of Chongming who had been appointed headman (*zong zhushou*) of Shengshan Island in 1906, took over as the Shengshan Branch Fishing Association's director.[12] Ling Pengcheng, another Chongming native who held an administrative position in a popular-education institute in another county, came back to his native place to serve as one of the Branch Fishing Association's directors, as well as an instructor in the Fishery Direction Office.[13]

Cage Fishing

In the early 1930s, the appearance of a new fishing technology in Zhoushan presented an unforeseen challenge for unofficial forms of regulation, as well as Jiangsu's recently founded fishery management organs. During the Republican period, the practice of using bamboo cages to catch cuttlefish, which had first originated in Guangdong and Fujian, spread north to the Wenzhou and Taizhou regions of southern Zhejiang. This fishing technique took advantage of cuttlefishes' habit of spawning on reefs covered with aquatic plants. Mistaking bamboo gear for these plants, cuttlefish swam into the cages and got trapped inside.

Cage fishing had two advantages over the conventional method of catching cuttlefish with nets. First, bamboo cages required considerably less capital and brought in a larger catch. On the average, fishermen could catch 80 *yuan* worth of cuttlefish in one season using nets, and 125 *yuan* worth using cages.[14] Second, bamboo cages fared better in inclement weather than did nets, which minimized the risk of losing the catch to winds and currents. These advantages made cage fishing a lucrative side-employment for households in Wenzhou and Taizhou that farmed most of the year and caught cuttlefish in the waters of their native regions in the spring and early summer.

The number of fishermen using cages began to grow in 1930, when Wenzhou migrants in the Zhoushan Archipelago brought in a large catch with this gear.[15] Cage fishing experienced an even larger upsurge in 1931, as flooding in southern Zhejiang deprived many farmers of their usual livelihood and coastal households turned to catching cuttlefish as an alternative source of income.[16] After growing numbers of cage fishers crowded fishing grounds in Wenzhou and Taizhou, many started to hire steamships and travel north to fish in the Shengsi Islands off of southern Jiangsu.[17] Sojourning cage fishers from Taizhou and Wenzhou constructed bamboo and grass shacks in the islands as temporary residences during the fishing season and returned to their native regions as soon as it came to an end.[18]

Despite its economic advantages, cage fishing did considerable damage to cuttlefish stocks. Cuttlefish released their eggs when they were caught, leaving them stuck to the sides of the cage. Many fishermen simply scraped off the cuttlefish eggs after pulling the cages out of the water. This destroyed the eggs and prevented them from hatching and growing into mature cuttlefish.

Fig. 11   Cuttlefish cages. From Li and Qu, *Zhongguo yuye shi.*

Since the eggs did not have the chance to develop, the extensive use of this fishing gear threatened to reduce cuttlefish populations.[19] However, long-term preservation of cuttlefish stocks mattered little to migrants from Wenzhou and Taizhou, for whom the profits obtained from cage fishing in the spring was a welcome source of additional income.

Of course, fishing with nets had a sizable impact on Zhoushan's marine ecosystem as well. Stow nets, as mentioned above, contributed to the depletion of stocks by catching large numbers of immature fish. Referring to the widespread practice of catching immature yellow croaker for sale as fertilizer, in the early 1930s one fishery expert went so far as to warn: "If this goes on, in several decades large yellow croaker may disappear from the ocean (*jueji yu haiyang*)!"[20] No evidence indicates that the local regulations enacted by net-using fishermen did anything to limit catches.[21] According to a 1935 investigation of stow-net fishing in the Shengsi Islands:

Since this kind of enterprise requires heavy capital and yields few profits, it is no longer easy to engage in. Moreover, because of an evil practice passed down through the years, fishermen who work on boats can set up stakes for nets of their own accord, and fishing households who own the boats cannot get involved. For this reason, many fishing households suffer financial losses and are gradually declining.[22]

No rules limited fishing activities, which resulted in declining returns to economic effort. In the mid-1930s, the number of stow-net boats in the Shengsi Islands had decreased by 80–90 percent compared to the 1890s.[23] As productivity declined and losses grew, fishermen turned away from stow-net fishing in waters close to the islands for more lucrative offshore fishing grounds.[24] Seen from this perspective, cage fishing added a new dimension to ongoing processes of environmental change.

The Politics of Fishery Management

Cage fishers were not solely responsible for ecological pressures on cuttlefish stocks, but their arrival in Zhoushan's fishing grounds led to diminished catches for boats that fished with nets. Tensions soon flared as the users of these two types of gear sought privileged access to common-pool marine resources and demanded restrictions on their opponents. As animosity toward the newcomers from Wenzhou and Taizhou grew, net fishers and their elite patrons called on the government to prohibit cage fishing. The initial impetus came from Yin county fishermen's Yongtai Lodge, which informed the Jiangsu provincial government that cage fishing had a harmful effect on cuttlefish stocks and urged its prohibition.

In the spring of 1930, Jiangsu's Agriculture and Mining Department petitioned the Nationalist government's Ministry of Industry to ban cage fishing on the basis of the Fishery Law, which gave official agencies power to prohibit activities that damaged the reproduction of fish stocks. After receiving the Ministry of Industry's approval, Jiangsu issued an order making the use of this new fishing gear illegal. The central government's Jiangsu-Zhejiang Fishery Affairs Bureau, at this point still subordinate to the Ministry of Finance, likewise endorsed the prohibition.[25] In the spring of 1931, the Yongtai Lodge urged Zhejiang province's Reconstruction Department to follow Jiangsu's example and force fishermen who used cages to switch to nets. Zhejiang responded by issuing an order banning cage fishing.

Jiangsu and Zhejiang's prohibitions drew protests from the elite leaders of Taizhou and Wenzhou cage-fishing groups, who retained the services of a lawyer from Taizhou to argue their case to the Nationalist government. In a complaint submitted to the Ministry of Industry in the summer of 1931, cage-fishing representatives admitted that all fishing activities could damage spawning in lakes but claimed that the size of fish runs in the ocean depended entirely on weather and wind conditions unaffected by human activities. For this reason, they argued, the Fishery Law's restrictions on gear deemed harmful to spawning did not apply to Zhejiang and Jiangsu's coastal fishing grounds. Instead, rather than restricting cage fishing, the provincial governments and the Jiangsu-Zhejiang Fishery Management Bureau (the official name of the Fishery Affairs Bureau after it was transferred to the Ministry of Industry) should promote the new technology as a means of increasing production and developing China's fishing industry.[26]

These claims do, in fact, have some validity. Fish, like all natural resources, are part of a complex and unpredictable environment that fluctuates at random, with no apparent relation to human intervention.[27] In the Zhoushan Archipelago, cuttlefish runs varied considerably depending on environmental fluctuations such as changes in weather and shifts in ocean currents.[28] The 1906 Maritime Customs trade returns drew attention to this phenomenon by noting that the success or failure of cuttlefish season in Zhoushan "depends largely on the weather."[29] Unlike yellow croaker and other species, cuttlefish production in coastal waters was inversely related to rainfall levels.[30] Hence, it is plausible that the rainy conditions that prevailed in 1931—a year of extreme flooding along the Yangzi River—contributed to the low cuttlefish catches of that year.[31]

But regardless of its merits, the Wenzhou and Taizhou fishermen's argument failed to convince the Ministry of Industry, which rejected the complaint on procedural grounds.[32] Fishery experts in the Ministry of Industry's Fisheries Office (Yuye ke) went even further, advising that because prohibiting cage fishing in no way conflicted with the Fishery Law, the government could simply disregard the complaint.[33]

Petitions that defended cage fishing in terms of social welfare gained a more favorable response. Representatives of the Wenzhou and Taizhou groups claimed that replacing cages with nets would cause serious financial losses for the residents of their native regions. Since cage fishers had already purchased their gear, it would be impossible for them to switch to nets so close to the start of cuttlefish season. Depriving Taizhou and Wenzhou natives of the income they gained from cage fishing would force them to turn to piracy. This reasoning won over Zhejiang, which ordered Dinghai and other coastal counties to permit the use of cages temporarily and wait until the following year to require fishermen to change to nets.[34] As a result, even though Jiangsu and the Ministry of Industry still banned cage fishing, Zhejiang changed its stance and let fishermen from Taizhou and Wenzhou use this type of gear.

To get around resistance to cage fishing in Dinghai county, the director of Taizhou's Wenling County Fishing Association, Bao Heng, worked out a temporary settlement between net and cage fishers. In typical fashion, the agreement allocated fishing territories around the islands to opposing native-place groups in order to prevent conflict. Efforts by fishermen from Taizhou and Wenzhou to end Jiangsu's prohibition on cage fishing gained assistance from the former Dinghai county magistrate, Zhang Yin, who had become director of the Chongming and Qidong County Government-Owned Polder Fields Bureau (Chongming jian Qidong shatian guanchan fenju).

Born in Taizhou's Linhai county, Zhang Yin graduated from a police academy in Japan during the final years of the Qing. After returning to China in the early 1910s, Zhang took a post as a police chief in his native Zhejiang. His success in subduing a bandit uprising in Sheng county brought him to the attention of Zhejiang provincial governor Qu Yingguang, who was also a native of

Linhai. With Qu's patronage, Zhang secured appointments as magistrate in Zhejiang's Dongyang, Linhai, and Dinghai counties. As noted in Chapter 3, during his tenure in Dinghai county during the early 1920s, Zhang had been instrumental in mediating feuds between rival fishing groups on Daishan Island. After Qu Yingguang became governor of Shandong, Zhang Yin went with him and acted as a magistrate in several counties in Shandong as well. Thanks to Zhang's political connections, he later gained an appointment as the head of the Wenzhou-Taizhou Polder Fields Management Office (Wen-Tai shatian guanlichu), which developed and then collected taxes on lands formed by the build up of silt in coastal areas. During the Nanjing decade, land developers who controlled polder-field bureaus were among the most indomitable "local bullies and evil gentry" who plagued Nationalist efforts to gain control of local government.[35] This lucrative post made Zhang Yin one of the wealthiest men in the Taizhou region.[36] Furthermore, Zhang's role in the administration of polder fields brought him into contact with the Wenling County Fishing Association's directors Chen Zhongxiu and Liu Boyu, who held posts alongside Zhang in land offices in Taizhou.[37]

Intense rivalry existed between the Taizhou migrants connected with Zhang Yin and the members of Chongming's Nationalist Party branch who had organized the Shengshan Branch Fishing Association. The troubles began when the Nationalist Party branch brought allegations of corruption against Zhang Yin's administration of the Chongming and Qidong County Government-Owned Polder Fields Bureau. To eliminate these malpractices, the party branch asked the Jiangsu provincial government to get rid of the Polder Fields Bureau and assign the management of polder land to Chongming county. The resulting recriminations led to an acrimonious string of lawsuits.[38] By the 1930s, the struggle between Chongming and Taizhou elites over income from polder lands broadened to contests for the control of profits from Zhoushan's fishing industry.

In the spring of 1932, Zhang Yin put skills honed in the polder business to good use and petitioned the Jiangsu-Zhejiang Fishery

Management Bureau to eliminate prohibitions on cage fishing enacted by Jiangsu and the central government.[39] Given the financial difficulties and corruption that plagued the Fishery Management Bureau from its founding, bribes undoubtedly accompanied Zhang's requests.[40] With this persuasion, the Fishery Management Bureau's director, Han Yougang, issued an order lifting all bans on cage fishing. According to Han Yougang, investigations undertaken by the Fishery Management Bureau found that cage fishing did no harm to cuttlefish spawning. Moreover, loss of income due to the cage-fishing prohibition would compel migrants from Taizhou and Wenzhou to turn to piracy. As an alternative, Han Yougang endorsed the Wenling Fishing Association's earlier agreement dividing up fishing grounds between net and cage fishers.[41]

Han's actions aroused vigorous opposition from the Shengshan Branch Fishing Association, which alleged that "fishing bullies" had coerced the Fishery Management Bureau into making this proclamation. The Fishing Association's leaders urged the Ministry of Industry to nullify the order and dismiss Han Yougang from office.[42] At the same time, prominent Ningbo elites appealed to the Nationalist regime on behalf of fishing groups from their home region. Zhuang Songfu, acting in his capacity as head of the Nanjing Ningbo Native-Place Association (Ningbo lü Jing tongxianghui), urged the central government to nullify the Fishery Management Bureau's proclamation.[43] The banker Yu Qiaqing, head of the Ningbo Native-Place Association in Shanghai, also pushed the government to protect the interests of net fishermen. The Shanghai Taizhou Native-Place Association, on the other hand, lobbied the Nationalist government to allow the use of cages.[44]

From the outset, the Ministry of Industry refused to approve the Fishery Management Bureau's decision to lift the ban on cage fishing since it had issued the order without first consulting Zhejiang and Jiangsu. The Ministry of Industry also made it clear that the Fishery Management Bureau could not alter earlier prohibitions on cage fishing without the ministry's permission.[45] But despite the Ministry of Industry's admonitions, the Fishery Management Bureau took no action to revoke its directive.

The director of Jiangsu's Department of Industry, He Yushu, insisted on keeping the prohibition on cage fishing in place, citing reports from the Jiangsu Provincial Fishery Experiment Station on the damage that cages did to cuttlefish stocks. Moreover, He pointed out, the Fishery Management Bureau's order violated the Fishery Law, which required appeals regarding Jiangsu's fishery regulations to be made directly to the province. For He Yushu, Jiangsu's ability to regulate its fisheries was a matter of provincial sovereignty. Each province had different "geography and customs," and government policy had to take "local peoples' substantive experience" (*gedi renmin liti jingying*) as its standard. Fishermen in Jiangsu's Shengsi Islands had always used nets, and cage fishing harmed cuttlefish breeding. Therefore, Jiangsu's prohibition on cages had to be upheld. All cage fishers from Wenzhou and Taizhou who came to Jiangsu needed to adhere to its laws by using nets or give up fishing altogether.[46]

For fishery experts in the Nationalist government, the cage-fishing controversy was a matter of conserving natural resources needed for the future development of China's fishing industry. This priority reflected the goal of increasing production by eliminating waste and promoting efficient exploitation. The Ministry of Industry handled the cuttlefish issue largely according to recommendations from the head of its fishery division, Li Shixiang. In Li's view, protecting cuttlefish stocks was necessary to promote a Chinese industry that competed against foreign goods. China imported large amounts of dried squid from Japan each year, and this made it essential to preserve the stocks that provided raw materials for domestically produced dried cuttlefish.[47] Since cage fishing damaged cuttlefish spawning, Li argued, it had to be banned in accordance with the Fishery Law. Based on Li Shixiang's interpretation, the Ministry of Industry instructed the Fishery Management Bureau to uphold Jiangsu's ban and issue an edict retracting its previous announcement permitting the use of cages.[48]

Even though the Ministry of Industry's decision favored Jiangsu, it did nothing to alter Zhejiang's lenient stance toward cage fishing. Zhejiang no longer prohibited this gear, and so the retraction of the Fishery Management Bureau's order had no effect on its policy. In early May, Zhejiang's Reconstruction Department ordered all coastal counties to follow the Fishery Management Bureau's plan to divide up fishing territories between net and cage-fishing groups. The Ministry of Industry could do nothing more than order Zhejiang to carry out investigations to determine if cage fishing depleted cuttlefish stocks in its coastal waters.[49] In this way, crosscutting administrative jurisdictions led to contradictory and inconsistent policies, which increased the potential for violent conflict over resources.

Fighting for Fish

Even after the Ministry of Industry nullified the Jiangsu-Zhejiang Fishery Management Bureau's order, cage fishers and their elite patrons found ways to gain access to fishing grounds off Jiangsu's Shengsi Islands. In April 1932 Taizhou migrants led by Ge Liquan, director of the Taizhou Fish Merchants Lodge, unsuccessfully tried to get official permission to organize a fishing lodge to protect the interests of cage fishers from Wenzhou and Taizhou who gathered in the Shengsi Islands during cuttlefish season.[50] Undaunted by this setback, Ge Liquan, together with one of Zhang Yin's functionaries in the Polder Fields Bureau named Chen Lepu and Chen's son Chen Xiangkui, paid off a subdivision of Zhejiang's coast guard to defend cage fishers who entered Jiangsu waters.[51] The Taizhou group's act of defiance was resisted by its rivals in Jiangsu's Fishery Direction Office and the Shengshan Branch Fishing Association.[52]

Under these circumstances, simmering tensions between net and cage fishers ignited into violence. Citing a letter from Li Shixiang to Jiangsu's Fishery Experiment Station rebuking the Fishery Management Bureau's decision to lift the ban on cage fishing, Jiangsu's Fishery Direction Office announced that the prohibition on cage fishing was still in effect and exhorted local fishermen to "unite their spirits and oppose this unlawful invasion."[53] Conflicts between the two groups culminated in a major feud near Huaniaoshan Island in early June. Several hundred well-armed net fishermen clashed with cage fishers and their supporters in Zhejiang's coast guard. Jiangsu responded by sending its own coast guard units to enforce the ban and put down the dispute. To seek refuge, the cage fishers boarded a steamship and hightailed it to Shanghai. When the dust settled, fighting had resulted in at least eight deaths and dozens of injuries.[54]

This battle acted as a catalyst for redefining Jiangsu and Zhejiang's coastal boundaries and clarifying jurisdiction over the Shengsi Islands. At Jiangsu's bequest, the Ministry of Industry and Ministry of the Interior convened a meeting between the provincial governments on July 5, 1932, to resolve the border issue. The fishery expert Hou Chaohai, who represented Jiangsu at the meeting, made it clear that the cuttlefish dispute had prompted these negotiations: "Fishers from Wenzhou and Taizhou in Zhejiang have caused problems that Zhejiang cannot solve, and the controversy has impinged on Jiangsu's territory. Jiangsu rightly issued a prohibition to support and protect spawning, and this led to the fishing dispute." Provincial boundaries had to be redrawn to stop disputes from occurring in the future. To facilitate administrative control, Hou proposed making the islands currently under Jiangsu's jurisdiction a special district, which the province would administer along with the central government.

Although Zhejiang's representatives claimed that they were merely passive participants in the meeting, they took the negotiations as an opportunity to lay claim to the entire Zhoushan Archipelago, including Jiangsu's Shengsi Islands. Placing all the islands under Zhejiang's control, they argued, would facilitate local administration and eliminate the possibility of further fishing disputes.[55]

The funding concerns of fishery experts employed by provincial government agencies factored into the territorial controversy as well. Jin Zhao, director of the Zhejiang Provincial Fishery

School, stressed that the Zhoushan Archipelago needed a unified administration to implement the program for fishing harbor development inspired by Sun Yat-sen's *Industrial Plan*. Nevertheless, the call to unify the archipelago by changing the jurisdiction of the Shengsi Islands did not derive simply from a desire to realize Sun's legacy. Fishery experts had hoped to designate Shengshan Island as the first site for a modern fishing port since the mid-1920s, and the Nationalist government's Four-Year Plan of 1933 specifically called for the construction of this facility.[56] Thus, fishery experts in the province that controlled the Shengsi Islands would become direct beneficiaries of the Nationalist government's economic development plans.

The fishery expert He Huiyu, who attended the negotiations in place of Li Shixiang, explained that the Ministry of Industry had no opinion on the border question.[57] However, as He Huiyu explained, effective fishery management in the Zhoushan Archipelago required clear administrative boundaries: "For example, if one island is under the jurisdiction of two provinces, then fishing territory must be divided between the coast guards from two areas. As they carry out their administrative duties, this can easily lead to conflicts over jurisdiction and all kinds of disputes may arise." Obviously, He Huiyu was referring to the jurisdictional conflicts that had emerged during the cuttlefish feud. Uniting the entire archipelago under the administration of a single province would avoid similar conflicts in the future.

The Ministry of the Interior took a different stance, arguing that "natural land formations" and "administrative convenience" rather than fishery management were the appropriate grounds for deciding the territorial question. With these considerations in mind, the Ministry of the Interior could approve placing all the islands under Zhejiang in principle. But it also recognized the practical difficulty of implementing this change. For this reason, the Ministry of the Interior wanted the provinces to come up with their own solution to the border problem.[58]

After much deliberation, the talks produced a two-point plan for clarifying the Zhejiang-Jiangsu boundary. The two provinces would place Huanglong Island under Zhejiang's administration, and Jiangsu would retain control over the islands of Tanxu and Baishan. According to the Ministry of the Interior's regulations for determining administrative boundaries, the two provinces had to meet and give their mutual endorsement before the central government could move forward with the settlement.[59] Unfortunately, the proposal satisfied neither Zhejiang nor Jiangsu, and the border question created by the cuttlefish feud remained unsettled.[60] As a result, competition for fishing grounds expanded from a local confrontation between rival native-place groups to an interbureaucratic conflict on the national scale.

Law and Ecological Conflict

As the 1932 cuttlefish season came to an end and conflicts between net and cage-fishing groups died down, the cuttlefish feud shifted to the courts. A surveyor in the Polder Fields Bureau accused Yu Zhenxiang, a director of the Shengshan Branch Fishing Association, of injuring him during the cuttlefish feud. Chongming county authorities hesitated to apprehend Yu Zhenxiang after Jiangsu's Fishery Direction Office circulated a draft edict from the provincial government relieving him of responsibility for the incident. With Yu Zhenxiang still at large, Zhang Yin and his associates in the Polder Fields Office obtained a warrant for Yu's arrest. He was apprehended by police in Shanghai during the following winter. The Taizhou group eventually failed to get revenge, since the Shengshan Branch Fishing Association and its allies in Chongming's Nationalist Party branch secured Yu Zhenxiang's release.[61]

The Taizhou and Wenzhou groups' lawyer also petitioned the central government's Administrative Yuan to appeal the Ministry of Industry's decision upholding Jiangsu's prohibition against cage fishing.[62] The Nationalist government again rejected the Taizhou and Wenzhou groups' arguments, since they had not petitioned the Ministry of Industry before bringing their case to the Administrative Yuan.[63] Still undaunted, the cage fishers' lawyer filed another complaint in early 1933, accusing Li Shixiang, Hou

Chaohai, and the Jiangsu Fishery Direction Office's director, Wang Liansheng, of taking bribes from net fishermen and inciting the bloody conflict at Huaniaoshan Island. This time, the complaint went so far as to ask the central government to hand the fishery experts over to Chongming county to face criminal charges.[64] The Ministry of Industry would have nothing of it, however, stating the cage fishers' representatives should direct any evidence for these allegations to the courts for judgment.[65]

The Limits of Scientific Fishery Policy

As opposing regional factions leveled accusations against each other in court, equally contentious debates raged over the legality of cage fishing. Divergent provincial and central government positions on cage fishing derived from the findings of fishery research institutes in Zhejiang and Jiangsu, which came to totally different conclusions about the gear's impact on cuttlefish stocks. Nationalist government fishery legislation required regulations based on objective scientific investigation. But with a complex, unpredictable environmental system like the Zhoushan Archipelago's fisheries, this unambiguous standard of evidence proved nearly impossible to obtain. Fishery experts in Jiangsu found that cage fishing damaged cuttlefish spawning and had led to the species' depletion. The Zhejiang Provincial Fishery School (Zhejiang shengli shuichanke zhiye xuexiao), by contrast, argued that no statistical data showed that cages did more damage than nets.[66] To settle this disagreement, the Ministry of Industry needed to conclusively demonstrate the effect of cage fishing on cuttlefish populations.

With this goal in mind, the Ministry of Industry organized a committee composed of fishery specialists, biologists, government officials, and fishing organizations to research cuttlefish spawning.[67] The Ministry of Industry delegated responsibility for this project to the Jiangsu-Zhejiang Area Fishery Improvement Commission (Jiang-Zhe qu yuye gaijin weiyuanhui), a new agency that it had established in February 1933. Along with fishery experts and Minister of Industry Chen Gongbo, the Fishery Improvement Commission included prominent fishing industry representatives such as the Yongfeng Lodge's director Zhang Shenzhi and the boss of Shanghai's Green Gang, Du Yuesheng, who also happened to be chairman of the Shanghai Frozen Fish Same-Trade Association.[68]

According to the fishery expert Hou Chaohai, the Ministry of Industry formed the Improvement Commission to centralize regulation of the fishing industry. The Nationalist government could not manage the Zhoushan Archipelago's fisheries according to conventional administrative divisions that applied on land. Plans for constructing modern fishing harbors and market facilities had to be "directly subject to central government plans." Therefore, as Hou explained, the Ministry of Industry established the Fishery Improvement Commission to prevent jurisdictional conflict, settle disputes between central and local authorities, and facilitate the interbureaucratic communication needed to successfully implement fishery policy.[69]

Funding for the new fishery management agency depended on revenues from the fishing industry. The Fishery Improvement Commission planned to collect a "fishery reconstruction fee" (*yuye jianshe fei*) of 2 percent on the price of fish sold by brokers. The Ministry of Industry pledged not to use these revenues for any purpose besides fishery development programs. Reconstruction fees would be "taken from fisheries and used for fisheries."[70] For fishery experts, obtaining a share of the revenues from this levy on fishing enterprises might remedy the budget shortages that had stifled their reform efforts since the 1920s.

The task of collecting reconstruction fees fell to Dai Yongtang, the secretary-general of Shanghai's Frozen Fish Same-Trade Association, who was appointed head of the Jiangsu-Zhejiang Region Fishery Reconstruction Fee Collection Office in Shanghai. Dai Yongtang selected Liu Jiting, director of the Shenjiamen and Dinghai branches of the Bank of China, to head another collection office in Shenjiamen.[71] Once this new fee was in place, the central government would abolish all informal levies on the fishing industry. For this reason, the Ministry of Industry claimed, the fishery reconstruction fee would actually reduce the tax burden on fishing enterprises. This argument did not convince fish brokers, who organized against the new levy under the leadership of the Shanghai banker Yu Qiaqing.[72] Eventually, in late 1934 the Ministry of Industry and fish brokers in Shanghai agreed to decrease the reconstruction fee to 1 percent. Income from this levy barely covered the Fishery Improvement Commission's expenses, forcing the central government to give up on this method of extracting revenues.[73]

Left without funding for research on cuttlefish spawning, the Fishery Improvement Commission settled on granting subsidies of 300 *yuan* to fishery schools in Zhejiang and Jiangsu to conduct cooperative investigations on the subject. Drawing on their findings, the Ministry of Industry would be able to decide the appropriate stance on cage fishing.[74] The Jiangsu Provincial Fishery School, the Jiangsu Provincial Fishery Experiment Station, the Jiangsu Fishery Direction Office, and the Zhejiang Provincial Fishery School made one futile attempt at joint research in the summer of 1933. From that point on, the provincial fishery institutes conducted all their investigations independently.

The Jiangsu Experiment Station reaffirmed that cage fishing was detrimental to cuttlefish spawning and would lead to depletion or even extinction if not prohibited. Jiangsu researchers acknowledged that nets also had the potential to harm reproduction of cuttlefish stocks. But they concluded that nets did far less harm than cages, since they did minimal damage to eggs. For this reason, net fishing did not present an immediate threat to cuttlefish stocks.[75]

The Zhejiang Provincial Fishery Experiment Station agreed that cuttlefish catches were in fact declining. Yet Zhejiang researchers stressed that this decline resulted from excessive fishing efforts by nets *and* cages. No evidence existed that one type of gear did more harm than the other. The government could solve the problem of declining stocks by eliminating the harmful practice of scraping eggs off cages, restricting all fishing activities at the height of spawning season, and encouraging research on artificial cuttlefish breeding.[76]

According to Zhejiang researchers, the cause of the cuttlefish feud was the conflict of interest (*lihai chongtu*) between net and cage-fishing groups. The dispute "had absolutely nothing to do with the problem of spawning."[77] Cage fishers occupied a portion of the limited number of spots where cuttlefish could be caught. This reduced the catches of net fishers who had previously monopolized the most productive fishing territories. Opposition to cage fishing among net fishermen derived not from a concern for preservation of cuttlefish stocks, but from the arrival of new competitors who cut into their economic returns.[78] To prevent further disputes, fishery experts in Zhejiang recommended that the government implement a system of clearly defined fishing rights. If net fishers and cage fishers possessed exclusive access to fishing grounds, Zhejiang researchers maintained, there would no longer be cause for conflict. Furthermore, the government needed to register all types of gear and gather statistics on cuttlefish catches from year to year.[79] Although the plan made sense on paper, Zhejiang's fishery experts never explained how they were to be implemented.

During the 1920s and 1930s, fishery researchers in Europe and North America engaged in contentious debates over the effects of harvesting activities on fish populations.[80] For this reason, it would be unfair to fault Zhejiang and Jiangsu researchers for their inability to agree on the factors that led to declining cuttlefish catches. But despite their avowed quest for impartial scientific information, funding considerations also influenced the research priorities of Chinese fishery experts. The Zhejiang Provincial Fishery Experiment Station's proposal for halting the decline in cuttlefish stocks, for example, requested 1,000 *yuan* a year from the Nationalist government to support research on stock-raising programs.[81] By becoming the lead agency in efforts to resolve the cuttlefish-spawning issue, provincial research institutes endeavored to gain financial backing from the central government.[82] This competition for budgetary support made it even more difficult for bureaucratic agencies to come to a consensus on the effects of cage fishing.

With the cage fishing issue unsettled, Jiangsu did not alter its prohibition against the gear. Zhejiang, for its part, refused to stop Taizhou and Wenzhou migrants from violating the neighboring province's policy and fishing with cages in Jiangsu waters.[83] Seeking to avoid further complications, the Ministry of Industry declined to intervene in the disagreement, because the question of cage fishing belonged "within the scope of local fishery administration."[84] Du Yuesheng made repeated efforts to arbitrate an agreement between net and cage-fishing groups, but these, too, ended in failure.[85] Sporadic conflicts between the fishing groups persisted until the outbreak of the Sino-Japanese War in 1937.[86]

Conclusion

At the most fundamental level, the cuttlefish feud resulted from the introduction of a new technology—cage fishing—that aggravated common-pool problems in the Zhoushan Archipelago during the early 1930s. Cage fishers from Wenzhou and Taizhou competed with net fishers for a finite body of resources, leading to decreasing returns to economic effort. At the same time, the environmental impact of cage fishing intensified pressures on cuttlefish stocks, leaving less for everyone. The presence of cage fishers complicated the thorny problem of allocating access to unevenly distributed natural resources, leading to disputes with net fishers over access to the most productive spots. Settling the conflicts sparked by cage fishing required coming up with new arrangements coordinating the use of Zhoushan's fisheries.[87] Yet the divergent interests of the parties that claimed access to these natural resources precluded the creation of new institutions governing their use.

After natural disasters disrupted the livelihood of many households from Taizhou and Wenzhou, they enthusiastically turned to cage fishing as a profitable new technology to supplement their income. For newcomers from peripheral regions of southern

Zhejiang, profiting from the exploitation of cuttlefish in the short term clearly mattered far more than long-term preservation of the resource. But since net fishers did not exploit marine resources in a sustainable fashion either, it makes little sense to attribute their opposition to cage fishing to concerns about the conservation of fish stocks. For net fishers, the crux of the problem was that cage fishers reduced the income they derived from Zhoushan's fishing grounds. From this perspective, the losses that would have resulted from a compromise with cage fishers outweighed any benefits net fishers stood to gain from an agreement. These differing priorities ruled out a cooperative solution to conflicts over the use of cuttlefish stocks. As a result, opposing native-place groups in Zhoushan devoted their energies to predatory and defensive activities.

By the early 1930s, a complicated web of state agencies with unclear jurisdictions vied for revenues collected from fishing enterprises. National, provincial, and local governments tried to maximize their fiscal income by getting a cut of the profits from Zhoushan's fisheries. Government agencies that benefited from fees levied on fishing groups were split between those allied with net fishers and those that supported cage fishers. The resulting political divisions increased the potential for conflict between regional groups. Moreover, Zhejiang and Jiangsu provinces based their policy positions on the findings of fishery experts who also had an interest in extending control over Zhoushan's fishing industry. For provincial fishery agencies, compromise on the cage fishing issue weakened their claims to the sizable revenues that came with jurisdiction over Zhoushan's fishing grounds.

During the Republican period, Chinese government legislation attempted to formulate rational and consistent guidelines for regulating marine resources. But in practice, the impossibility of obtaining the unambiguous scientific evidence demanded by the government's fishery laws delayed action on fishing disputes and complicated resolution of conflicts. As a consequence, this legislation entangled fishing disputes that in the past had been handled through local social and political networks in the complex web of the modern Chinese state's administrative bureaucracy. Official handling of the cuttlefish feud cast the gap between the Nationalist regime's centralizing aspirations and its limited administrative capacity in sharp relief.

During the 1930s, as the scale of fishery regulation moved from local networks to the Nationalist bureaucracy, the task of settling conflicts over Zhoushan's fishing grounds grew all the more difficult. Bureaucratic intervention complicated the resolution of resource disputes just as intensifying demands on fish stocks increased the potential for conflict. Arrangements that native-place groups relied on to cope with competition for finite common-pool resources no longer worked; the new ones that state agencies advocated made the problem worse. The unceasing battles between net and cage fishers during the 1930s reflected this unfortunate convergence of nature, society, and politics.

SIX

Fishing Wars III: The Zhejiang-Jiangsu Border Conflict, 1935–1945

The environment is always unpredictable, and at times its fluctuations have political repercussions. During the mid-1930s, unexpected shifts in the geographical position of the Zhoushan Archipelago's main fishing grounds led fishermen from Zhejiang to migrate en masse to Jiangsu's Shengsi Islands. This environmental change made collection of fishing taxes a major point of contention between local governments in Zhejiang and Jiangsu, as well as among the fishing lodge leaders they relied on to collect these fees. Even as Japan's military presence loomed in Manchuria and North China during the late 1930s, the two provinces over which the Nationalist regime claimed firmest control clashed over offshore fishing grounds. With state agencies struggling to extract greater revenues from Zhoushan's fishing industry, interbureaucratic conflict grew more intense. Under these circumstances, the problem of conserving fish stocks raised during the cuttlefish feud was no longer an official priority. Instead, state-led programs to expand fishery output and generate greater tax revenues through more efficient exploitation of marine resources persisted until the outbreak of full-scale war between China and Japan brought nearly all fishing in the Zhoushan region to a halt during the 1940s.

The Jiangsu Business Tax

By the early 1930s, the opening of offshore fishing grounds by
Zhejiang fishermen had transformed the Shengsi Islands, located
at the northernmost reaches of the Zhoushan Archipelago within
Jiangsu's provincial borders, into a burgeoning fishing center. The
potential revenues to be gained from taxing Shengsi's fishing enterprises did not escape the attention of officials in Jiangsu province. In January 1935 the Chongming Business Tax Collection Office (Chongming yingyeshui zhengshouchu), a fiscal organ directly
subordinate to the Jiangsu province's Department of Finance, began to collect a short-term business tax (*linshi yingyeshui*) of 0.5
percent from fish brokerages in the Shengsi Islands.[1] Jiangsu estimated this levy could bring in several thousand *yuan* in revenue
each year.[2]

Although the central government in Nanjing had waived all
fishing taxes to protect domestic producers against Japanese
competition four years earlier, Jiangsu's Department of Finance
tried to justify its levy on fishing enterprises as a tax on business
firms (*an hang wei keshui*), which made it different from a tax on
goods (*feng wu keshui*) such as fish. For this reason, Jiangsu officials ordered fish brokers in the Shengsi Islands to withhold a tax
of 0.5 percent of the value of goods on all business transactions.[3]
This method of delegating tax collection to brokers paralleled
Jiangsu's policy of allowing local authorities to farm taxes instead
of collecting them according to formal bureaucratic procedures.[4]
Pointing to the difficulty of collecting taxes in the remote
Shengsi Islands, the provincial government set aside 30 percent of
the proceeds from the business tax as compensation for the personnel who collected this tax from fish brokers.[5]

Before long, Jiangsu's new business tax ignited objections from
fishing industry interests in Zhejiang. Fishing lodges and other
organizations wired the Nationalist regime and the Zhejiang provincial government, urging them to convince Jiangsu to halt collection of the business tax from fish brokerages in the Shengsi

Islands. Fishing lodge organizations claimed that the levy violated the central government's policy of waiving taxes on fishing enterprises.[6] After Jiangsu rejected appeals to eliminate the business tax, fishing organizations in Shenjiamen assembled on May 4, 1935, to devise a response. Leaders from numerous Ningbo-region fishing lodges, as well as the Shengshan Sojourners Lodge (Lü Sheng gongsuo), the Yin County Fishing Association, the Eastern Yin County Fishing Cooperative (Yin dong waihai yuye hezuoshe), the Shenjiamen Pair Fishing Same-Trade Association (Ding Shen dui yuye gonghui), and the Shenjiamen Fish Brokers Same-Trade Association (Ding Shen yuzhan gonghui), attended the meeting. The organizations announced that since fishing was fundamentally different from other commercial activities, Jiangsu could not levy taxes on their trade. All fish brokers in Shengshan were ordered to refuse to collect the business tax. Both the Shengshan Sojourners Lodge and the Shenjiamen Fish Brokers Same-Trade Association pledged to contribute 100 *yuan* apiece to fund efforts to abolish the commercial tax, and large pair boats would pay a fee of 5 *jiao* for the same purpose.[7]

Fishing organizations selected the Shenjiamen Pair Fishing Same-Trade Association's directors Lou Yuren and He Zhizhen to travel to Chongming in July and demand the elimination of the commercial tax. But after hearing their complaint, Chongming officials arrested He Zhizhen on charges of inciting tax resistance.[8] When He Zhizhen was released after 25 days of imprisonment, he promptly filed a lawsuit against personnel in the Chongming Business Tax Collection Office, accusing them of dereliction of duty, fraud, and illegal collection of fishing taxes.[9] Chongming authorities placed fish brokers who refused to pay the business tax under arrest for tax resistance as well, but they were eventually released.[10] Seeking to quell the tax dispute, the Ministry of Finance issued an order stating that collecting the business tax from fishermen (*yumin*) violated central government orders, but levying the tax on fish merchants (*yushang*) did not. Without decisive central government intervention, the controversy went on unabated. The Chongming Business Tax Collection Office pressed fish brokers for the tax; merchants backed by fishing organizations refused to pay.[11]

Ward Administration in the Shengsi Islands

The sizable profits generated by the growth of fishing activities in the Shengsi Islands were also coveted by Chongming county, which held administrative jurisdiction over the booming fishing center. Jiangsu's business tax was an impediment to Chongming's own efforts to gain revenues from Shengsi's fishing enterprises. Hence, vigorous objections to the business tax came from Chongming's Nationalist Party branch, which called on Jiangsu to stop collecting it from fish brokers.[12]

The steering committee of Chongming's Nationalist Party branch had made a stab at exerting tighter bureaucratic control over Shengsi's fisheries in January 1932, when it proposed a sub-county ward (*qu*) encompassing the islands. Wards were originally intended to be the basic unit of local self-government in Republican China, but by the 1930s they had become administrative organs under the supervision of county governments.[13] These sub-county bureaucratic units carried out population registration, land investigation, and policing. Even more important, ward administrations took responsibility for taxation at the local level.[14]

Insufficient finances delayed establishment of a ward administration in the Shengsi Islands for several years. Reform of local administration did not get off the ground until the spring of 1934, when Chongming county deputed personnel to investigate local conditions and set up ward offices on Shengshan Island. Chongming county's Fifth Ward, which administered all the Shengsi Islands, was officially established that May.[15] Ward leaders came from the same elite groups who staffed the fishery offices that Jiangsu had set up in the Shengsi Islands during the early 1930s to counter incursions by Japan's fishing fleet. Ling Pengcheng, who had previously been director of the Shengshan Branch Fishing

Association and a member of the Fishery Direction Office, took over leadership of subcounty administration in the Shengsi Islands as ward chief (*quzhang*).[16]

Local political factions in Shengsi paralleled the divisions between Chongming elites and sojourners from Wenzhou and Taizhou that came to a head during the cuttlefish feud. From its inception, the ward administration challenged the dominance of Taizhou natives like Zhang Yin, with whom Chongming's Nationalist Party branch fought over revenues from polder lands and fees collected from fishing boats. From the perspective of Chongming elites, the Shengsi Islands' fisheries had been "taken over" by these Taizhou migrants.[17] As one ward official put it:

Cage fishing, according to the results of academic research, is especially unfavorable to spawning. Wenzhou and Taizhou people advocate cage fishing, and the county party branch and the fishing association oppose it. The director of the Chongming and Qidong [Government-Owned] Polder Fields Bureau, Zhang Yin, is a native of Linhai [in Taizhou]. He is greedy and avaricious, and comrades in the fishing association also oppose him. Chongming people have taken this righteous and just stance for the residents of the islands of Shengshan and Sijiao. . . . It was Chongming people who first advocated creating a ward in the Shengsi Islands, and an upright man with fishery experience is the ward chief.[18]

In the spring of 1934, the ward administration started to stringently enforce Jiangsu's ban on fishing for cuttlefish with cages in waters around Shengshan and Huaniaoshan Island. Ward leaders made further inroads against their rivals from southern Zhejiang by dissolving the Chongming Taizhou Native Place Association (Taizhou lü Chong tongxianghui) and the Chongming Wenzhou Native Place Association (Wenzhou lü Chong tongxianghui), which sojourning elites from southern Zhejiang had set up to organize their compatriots.[19] By eliminating these native-place associations, the local government in the Shengsi Islands tried to displace sojourning elites who competed with it for control of profits from the fishery.

Fishing Protection and Fishing Fees

In the mid-1930s, many regional fishing lodges from northern Zhejiang hired coast guard units to assist their mutual-defense forces and collected "protection fees" (*baohufei*) from boats to cover this expense.[20] The chief of the Zhejiang coast guard gave fishing lodges permission to form protective convoys and granted official commissions to lodge leaders in charge of these mutual-defense forces.[21] But responsibility for overseeing defensive arrangements was actually in the hands of the Fish Brokers Lodge in Shenjiamen. The lodge's directors Chen Mansheng and Zhu Yunshui rented a steamship to protect fishing boats. Zhu "requested" that the commander of a regiment of Zhejiang's coast guard station an inspector (*xunguan*) and twelve officers on the vessel, which also flew the coast guard's flag. The Renhe Lodge, formed by fishermen from Yin county, which had Zhu Yunshui as its director, paid the coast guard 200 *yuan* in silver dollars every month for its services.[22] The Renhe Lodge covered the convoy's expenses by charging every fishing boat a fee of 16 *yuan*, which was collected by the Fish Brokers Lodge in Shenjiamen. The Fujian group's Bamin Lodge and numerous other lodge organizations from Zhejiang also collected protection fees from fishing boats. Taken as a whole, these exactions amounted to 50,000 *yuan* each year.[23]

Taking fees collected from fishing boats out of the hands of Zhejiang elites was particularly important for Shengsi's ward administration because, like the Jiangsu provincial government, it needed the revenues to remedy its fiscal shortages. The ward administration's income of 28,150 *yuan* met only 10 percent of its expenses.[24] To raise income, the ward planned to collect a "public welfare levy" (*gongyi juan*) from fishing boats affiliated with sojourning groups (*ke bang*) from Taizhou and Fujian. The number of these boats had increased dramatically with the growth of Shengsi's fisheries since the 1920s. By collecting the fees, the ward estimated that it could bring in an additional 4,000 *yuan* annually.[25]

Prior to the ward's establishment, all government revenues in the Shengsi Islands were provisional levies (*tanpai*) that local authorities imposed on fishing boats on a temporary basis whenever they needed funds. These informal levies existed outside the regular budgets of local governments, making them difficult to regulate.[26] Local militias in the Shengsi Islands, which were under the command of gentry who lived on the islands and acted as the heads of townships and municipalities (*xiangbaozhang*), monopolized collection of these fees.[27] An investigation of local finances in the Shengsi Islands by the ward administration described the influence of these elites in local society: "The headmen (*zhushou*) of each island are their leaders, and now they serve as heads of townships and villages. Island residents and government officials view them as the local gentry, and for this reason the headmen associate with officials and dominate village government, controlling everything according to their own authority." The local government had a hard time taking militia organizations and local finances out of the hands of these firmly entrenched elites.[28] With the ward's tenuous hold over the Shengsi Islands, it had no choice but to forfeit responsibility for tax collection to these local powerholders. To levy this fee on fishing enterprises, the ward relied on the same elites who led the Shengshan Branch Fishing Association. In 1935 the ward deputed Yang Youcai, the Shengshan Island headman who held a position as one of the Shengshan Branch Fishing Association's directors, and other local headmen to collect fees from migrant fishing boats.[29]

The Ningbo Fishery Police

In Zhejiang, as in Jiangsu, local governments tried to stake a claim to revenues created by the expansion of Zhoushan's fisheries. As the ward moved to levy taxes on fishing enterprises in the mid-1930s, Zhejiang moved to assert control over fees that native-place lodges collected from fishing boats. Since the late 1920s, the Nationalist government had granted provincial governments the power to collect registration fees from fishing boats and supervise mutual-defense organizations. To take advantage of this potential income, in June 1934 the Zhejiang authorities sent the province's Fifth Special Area administrative supervisor (*di wu tequ xingzheng ducha zhuanyuan*), Zhao Cisheng, to investigate the collection of protection fees by fishing lodges that had not registered with the government as required by the Fishing Association Law.[30] Initially appointed by the Nationalists in certain areas of China during the "bandit-suppression" campaigns to eradicate the Chinese Communist Party in 1932, special administrative supervisors were in charge of internal security at a level of governance between the county and the province.[31]

In May 1935, Zhao Cisheng organized the Ningbo Fishery Police Bureau (Ningbo yuye jingchaju) to exert bureaucratic control over revenues from Zhoushan's fisheries. According to Zhao, a native of Fenghua county, the Fishery Police originated from his most famous compatriot. "Motivation for establishing the Ningbo Fishery Police Bureau came from Chiang Kai-shek's relief plans for Fenghua's Tongzhao and Xifeng fishermen."[32] The Fenghua fishermen who fought in the 1911 Revolution had supposedly impressed Chiang Kai-shek. Available sources, however, offer no verification for Zhao Cisheng's claim. What is clear is that invoking this connection with Chiang lent the Fishery Police an air of authority and legitimacy.

The Ningbo Fishery Police made their headquarters in the Fifth Special Area administrative supervisor's offices, and Zhao Cisheng took over as chief. The Fishery Police Regulations (*Yuye jingcha guicheng*) that the Nationalist government enacted in 1931 granted these special public security forces substantial authority in fishery management. According to the regulations, fishery police maintained safety and order in fishing areas, mediated disputes over fishing territories, prevented foreign encroachments on fishing grounds, and enforced policies intended to protect the reproduction of fish species.[33] Initially, the Ningbo Fishery Police had no funding and had to rely on loans that Zhao Cisheng obtained from a native bank. Zhejiang later approved the collection of registration fees from fishing boats to meet the Ningbo Fishery Police's expenses.[34]

As in Jiangsu, Zhejiang's local governments lacked the capacity to bring in these revenues through bureaucratic channels. For this reason, the Fishery Police had to enlist fishing lodge leaders in the collection of registration fees. Zhao Cisheng called on Liu Jiting, manager of the Dinghai branch of the Bank of China and director of the Shenjiamen Fish Brokers Lodge, to organize a management committee in Shenjiamen and come up with a strategy for underwriting the Fishery Police's finances. In April 1935 Liu gathered leaders from various native-place groups at the Shenjiamen Chamber of Commerce to discuss this task.[35] Zhuang Songfu, who had become Zhejiang's self-government preparatory commissioner (*zizhi choubei weiyuan*), traveled to Shenjiamen to participate in the meeting.[36] According to its draft regulations, the management committee consisted of representatives appointed by fishing lodges and other fishing organizations from the Ningbo region. Every month the management committee would pay the Ningbo Fishery Police's expenses and inspect its accounts. The committee would also report the amount of fees collected from fishing boats by lodges and other fishing organizations to the Fishery Police and make this information public.[37]

The Fishery Police's registration offices at the Shenjiamen and Daishan public security bureaus (*gongan fenju*) collected registration fees of between 3 and 8 *yuan* from each fishing boat. In return, boats from the Ningbo region received flags issued by the Fishery Police, with different colors given to each native-place group (*bang*). Fishing lodges distributed these flags and withheld the necessary fees from boats.[38] To facilitate its effort to administer and tax Zhoushan's fisheries, the Fishery Police also instated the *baojia* mutual security system in Shenjiamen, Daishan, Qushan, and other fishing centers. Local control plans again relied on fishing lodges' elite leaders, who compiled these mutual security registers.[39]

Environmental Changes in the Mid-1930s

Over the course of the Republican period, fishing boats in the Zhoushan region had ventured progressively farther from shore in their search for more productive fishing grounds. According to a 1935 article in the newspaper *Dinghai Zhou bao*, the seas off Shenjiamen and the waters around Daishan Island had once produced large amounts of yellow croaker. Later, "because of the extension of steamship navigation, the fish have been driven into offshore waters. Hence, fishing grounds have shifted progressively farther from shore." There is no way to know if coastal shipping actually played a part in the shift of fishing grounds into offshore waters, but this additional stress on the marine environment may have contributed to the trend.

The article went on to explain that Shengshan Island's fishing industry had started to expand only after the Jiangsu-Zhejiang Fishing Company's steamship *Fuhai* discovered abundant yellow croaker stocks in nearby waters. In the early 1930s, only a little over ten years since their initial discovery, fishing grounds off Shengshan already showed signs of declining catches. During the 1934 fishing season, large pair boats found abundant schools of yellow croaker in waters more than 30 *li* to the east of Shengshan Island.[40] This sudden change in the position of yellow croaker fishing grounds illustrates the complex, unpredictable character of the marine environment.[41] Again, there is no way to determine the exact combination of factors that led to this change, but it had major economic and political consequences. As fishing boats flocked to the new offshore fishing grounds, utter silence fell over formerly bustling fishing centers like Changtushan Island, where Zhejiang had once intended to construct a modern fishing harbor in accordance with Sun Yat-sen's *Industrial Plan*. In 1935, as the center of fishing activities moved north to waters far away from Changtushan Island, another news article worried, "The shift in fishing grounds will spell the death of Changtu Harbor's fishing industry."[42]

Fishing boats came back to Shenjiamen with abundant catches from newly opened fishing grounds off Shengshan Island. Shenjiamen's Pair Fishing Same-Trade Association (Dui yuye gonghui) saw the fishing grounds as a boon for the future of the fishing industry. The Same-Trade Association hired a steamship to survey the fishing territories, draw up maps, and distribute this information to its members so they could take full advantage of it during the following fishing season.[43] With its boats moving into these fishing grounds in ever-greater numbers, in April 1936 the Shenjiamen Pair Fishing Same-Trade Association set up temporary offices to handle its affairs on Shengshan Island during fishing season. Fearing the complications that might arise if it opened an office in another province, the Same-Trade Association petitioned the Dinghai county government for assistance. The association urged Dinghai county to ask Zhejiang's provincial government to send communications to Jiangsu province and make sure Chongming county would guarantee its protection.[44]

As they moved into more distant waters, most of the boats from which the Ningbo Fishery Police collected registration fees came from Zhejiang but actually fished in Jiangsu waters. When the Ningbo Fishery Police's steamship arrived in Shenjiamen in November 1935, it was met by a delegation representing local fishing organizations. The patrol vessel followed Zhoushan's fishing fleet as it sailed to Jiangsu's Shengsi Islands for the fishing season.[45] Acting in his capacity as head of the Fishery Police's managing committee, the head of the Yin County Fishing Association, Shi Meiheng, oversaw collection of fees from fishing boats. Leaders of other Zhejiang fishing groups, as well as the director of Fujianese fishermen's Bamin Lodge, Liu Huafang, came along as the managing committee's administrative staff. The Fishery Police stationed patrol vessels at Shengshan Island to collect fees from fishing boats from Zhejiang and to distribute registration flags to them.[46]

Interprovincial Competition for Fishing Fees

While Zhejiang enlisted fishing lodge leaders to bring in revenues, Chongming's Fifth Ward carried out competing plans to collect registration fees from fishing boats that migrated seasonally from Zhejiang to Jiangsu's Shengsi Islands. Based on Jiangsu's guidelines for compiling *baojia* mutual security registers for fishing boats, in 1935 the ward administration required all "outside" fishing boats that berthed in the Shengsi Islands for more than six hours to register and obtain a license from the ward's boat inspection office. By making sure that these fishing boats registered with the ward, the subcounty administration could charge them fees. The ward further challenged the mutual-defense forces maintained by Zhejiang fishermen's native-place organizations, stating that any protective convoys that did not have official authorization from local governments in Jiangsu could not land at ports in the Shengsi Islands. The ward's public security forces would punish any boats that violated these regulations.[47]

Zhejiang's and Jiangsu's competing claims to profits from fishing enterprises in the Shengsi Islands also led to tensions between the local elites who collected these revenues for local governments. At Xiaoyangshan Island, the ward entrusted the local township headman Zhang Guifang with levying fees on sojourning fishing boats. In the summer of 1935, with help from the commander of the local defense forces, Zhang demanded the Fifth Ward's public welfare levy from Fujianese fishing boats that landed at Xiaoyangshan. On top of the "public levy" (*gong chuan*) that the Fujianese fishing group normally paid to Xiaoyangshan's local defense forces, Zhang required that migrant fish merchants from Fujian hand over another 4 *yuan* for each fishing boat and 10 *yuan* for every fish brokerage.

When boats from Fujian sailed from Shenjiamen to Xiaoyangshan during the fishing season for yellow croaker in June and July, Fujianese fish brokers and their Bamin Lodge opened shop on the island as well.[48] After fish brokers from the Bamin Lodge went to the township government offices to protest Zhang Guifang's collection of the Fifth Ward's new levy, the commander of the local defense forces took one of them captive and refused to release him until they paid the necessary fees. Liu Huafang, a Fujianese fish broker in Shenjiamen who was one of the Bamin Lodge's directors, asked the Nationalist government to punish Zhang Guifang for illegally demanding protection fees from fishing boats.[49] Liu Huafang had ulterior motives in opposing the ward's levy, since he, too, collected registration fees from fishing boats as a member of the Ningbo Fishery Police's administrative staff.

Disputes between Jiangsu and Zhejiang over fishing fees escalated in December 1935. Fishing lodge leaders connected with the Ningbo Fishery Police set up a registration agency (*dengji tongxunchu*) at a Chinese medicine shop on Shengshan Island. All fishing boats that went ashore or came to collect money from fish brokers had to pay the office the required registration fees. Fishermen quickly brought the Ningbo Fishery Police's demands to the attention of the Shengshan Branch Fishing Association. On December 19 a mob of fishermen whose boats had been confiscated for failure to pay the Fishery Police's registration fees captured a member of the office's staff and took him to the Fifth Ward public security force's headquarters in Shengshan's Empress of Heaven Temple. The ward administration also rounded up the Bamin Lodge's leader Liu Huafang and several other members of the Fishery Police's staff. The ward headman Ling Pengcheng demanded that Shi Meiheng come as well, but Shi had gone back to the Fishery Police's steamship and refused to come ashore for questioning. On the following day, fishermen reported that Shi Meiheng had gone into hiding at a fish brokerage in Shengshan, and the local defense forces went to place him under arrest. After capturing Shi Meiheng, the ward authorities forced Shi and his colleagues to take them to the fish brokerages where they stowed their unissued flags and the customs registers they confiscated from fishing boats. The ward leaders handed Shi Meiheng over to Chongming county for punishment; after a month in jail he was finally released.[50]

Territorial Disputes

With local governments in Jiangsu demanding fees from fishing enterprises, fishing lodges advocated giving Zhejiang control over the Shengsi Islands. Impetus for this change came from none other than Shi Meiheng. In the spring of 1935, Shi asked Zhao Cisheng to petition the Nationalist government to put Shengshan Island under Zhejiang's jurisdiction to unify authority and prevent additional disputes. Fishermen from Zhejiang and Shengshan had a profound connection, Shi maintained, and keeping the island under Jiangsu's administration was at odds with reality. Based on Shi's proposal, Zhao Cisheng wired Chiang Kai-shek asking him to order Zhejiang and Jiangsu to amend the boundary between the two provinces and place Shengshan under Zhejiang's control.[51] In June 1935, the Zhejiang regional associations in Shanghai and fishing organizations in Ningbo and Dinghai took their demands even further, calling on Chiang Kai-shek and the Ministry of the Interior to transfer *all* the Shengsi Islands to Zhejiang.[52]

Native-place associations from Zhejiang pointed to the environmental changes that had taken place in the Zhoushan Archipelago's waters to back their territorial claims. In their view, the development of the Shengsi Islands derived from shifts in the location of fishing grounds and the corresponding movement of Zhejiang fishing boats into waters farther from shore:

In recent years, because the fishing grounds have moved east, our Zhejiang fishing boats have also gradually advanced east as far as the Sheshan Sea. Because they could use Shengshan Island's ports to avoid the wind and waves, they settled there permanently. Not until several decades ago did the island begin to multiply in abundance as it has today.[53]

Before fishing boats from Zhejiang arrived, Shengshan was nothing more than a "deserted island." Only after a fish market developed did the island gradually begin to prosper and turn into "a small city in the sea." Pioneering fishermen from Zhejiang who moved to the Shengsi Islands as they explored offshore fishing grounds deserved credit for opening up this island frontier.[54] In the mid-1930s, the vast majority of the boats that fished in the Shengsi Islands still came from Zhejiang. In addition, Shengsi's credit and money markets centered on native banks in Ningbo and Dinghai. According to these claims, the islands grew in tandem with their connections to Zhejiang.

Conveniently forgetting the hostilities between Ningbo and Taizhou groups that had flared during the cuttlefish dispute a few years earlier, petitions to change the Shengsi Islands' administrative jurisdiction chose to stress a shared Zhejiang identity. Ningbo natives had assimilated the small portion of the population of the islands that hailed from Taizhou, and their customs and language had already become "Ningbo-ized" (*Ninghua*). Zhejiang groups claimed that the only Jiangsu natives in the islands were a few official functionaries, and people from Zhejiang never intermarried with them.[55] Shi Meiheng, in another petition to Chiang Kai-shek, pointed to differences in regional dialect to justify his claim that the islands belonged to Zhejiang rather than Jiangsu.

The dialects of Shengshan Island and Chongming county are totally different. One is Zhejiang pronunciation, and one is Jiangsu pronunciation. There is a clear divide and an obvious difference. North of the Huai River, a tangerine (*ju*) is called a wild orange (*zhi*). The local character makes it so (*diqi shiran*). This case is just like that. The islands were originally Zhejiang territory. It has been many years since they changed to Jiangsu, but their language has not been Jiangsu-ized (*Suhua*). From this you can see that there is a natural boundary that divides them.

Because the residents of the Shengsi Islands and the natives of Chongming county spoke different dialects, it was only "natural" to put the islands under Zhejiang's administration. Otherwise, the "customs and human feelings" of officials and local people would not be in harmony, and the two groups could not interact. Given this irreconcilable divide, "government administration will lead to persistent mutual conflict."[56]

Zhejiang groups stressed that administrative jurisdiction no longer corresponded to social and political reality. Although the

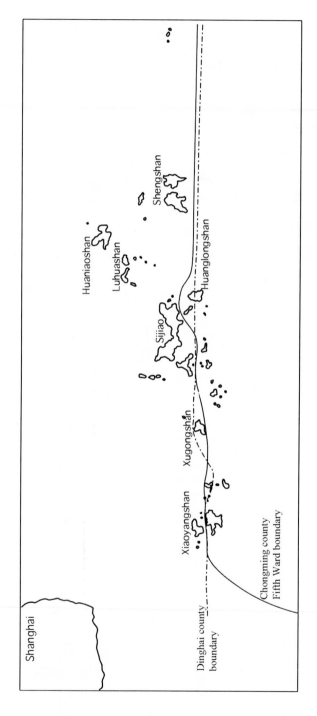

Map 3   Disputed boundaries in the Shengsi Islands.

Shengsi Islands fell within Jiangsu's provincial boundaries, social groups and officials from Zhejiang actually maintained law and order. Defensive convoys organized by Zhejiang's fishing organizations and patrol vessels from Dinghai county plied the waters around the Shengsi Islands during the fishing season to protect against piracy. Chongming paid no attention to these matters. Because of their distance from the Chongming county seat, Jiangsu authorities did not have any real influence on the Shengsi Islands. To the extent that Chongming administered the islands, it did so poorly. The financial accounts of local defense forces and ward administration offices were not open, and taxpayers could not consult them. To make matters worse, the "fishing tax in changed form" levied by Jiangsu's Business Tax Collection Office placed an unbearable burden on the fishing population. Shengshan Island had already surpassed Shenjiamen as the Zhoushan Archipelago's pre-eminent fishing port, and Zhejiang groups argued that good government was needed to plan for its future development.[57]

Chongming's Nationalist Party branch, Shengsi's ward administration, and Jiangsu's fishery management offices strongly opposed this effort to extend Zhejiang's administrative control over the islands. This resistance gained support from Jiangsu notables, such as the Shanghai gang leader Du Yuesheng, who supported Jiangsu's jurisdiction over the Shengsi Islands.[58] To back up their claim, the ward administration downplayed the role of Zhejiang natives in the development of Shengsi's fisheries. Instead, they attributed the growth of fishing enterprises in the islands to environmental fluctuations that had occurred independent of human action: "The shift to prosperity and the fishing grounds' move to the east were natural trends and have absolutely nothing to do with jurisdiction."[59] Even though most of the fishing population in the Shengsi Islands hailed from Zhejiang, Shanghai was the most important market for Shengsi's fish products and had replaced Ningbo as the industry's economic center of gravity.

Fishery experts from Jiangsu argued that their province had stood at the forefront of official efforts to develop the Shengsi

Islands' fisheries in the face of foreign competition. The head of the Jiangsu Provincial Fishery School, Zhang Yulu, traced the province's contribution to Shengsi's fisheries back to Zhang Jian's founding of the Jiangsu-Zhejiang Fishing Company in the final years of the Qing. More recently, Jiangsu had strengthened the fishing industry by founding the Fishery Direction Office in the Shengsi Islands and setting up Chongming's Fifth Ward. If Jiangsu were given sufficient time to implement its fishery development plans, it would end Japan's exploitation of fishing grounds in waters off the Shengsi Islands.[60] According to Jiangsu's fishery experts, Zhejiang needed to devote more energy to improving its own fishing industry and less to interfering in Jiangsu's internal affairs. Zhejiang already had the fishing port of Shenjiamen, and Sun Yat-sen's *Industrial Plan* included a modern fishing harbor on Zhejiang's Changtushan Island. Because Jiangsu possessed only the Shengsi Islands, depriving the province of this territory would do irreparable damage to Jiangsu's fishing industry. Only through proper division of labor and cooperation between the two provinces could China accomplish the "reconstruction" of its fisheries and resist further Japanese encroachments.[61]

While Jiangsu was protecting the Shengsi Islands' fisheries, Jiangsu groups claimed, disgruntled "vagrants" from Zhejiang's Taizhou and Wenzhou regions were agitating to change the islands' provincial designation so they could overturn Jiangsu's ban on cage fishing and regain their ability to collect fees from fishing boats.[62] The establishment of ward administration in the Shengsi Islands had already cut into Zhejiang fishing organizations' control of these fees. By trying to give their province control of the Shengsi Islands, Zhejiang natives simply wanted to expand the territories in which they could collect fees from fishing boats. Jiangsu's coast guard forces regularly sent vessels to waters around the Shengsi Islands to protect fishing boats and, at least according to Jiangsu's claims, never demanded fees for this service.[63]

Despite Jiangsu's vigorous opposition, Chiang Kai-shek initially endorsed Zhao Cisheng's proposal to reassign Shengshan Island to Zhejiang's Dinghai county. However, Chiang's decision ran into opposition from the Nationalist government's Ministry of the Interior, which took a different stance on the question. According to the Ministry of the Interior, the meetings that it held with the Ministry of Industry and Zhejiang and Jiangsu provinces in 1932 during the cuttlefish feud had already settled the border issue. The negotiations had assigned Huanglong Island to Zhejiang and placed the islands of Tanxu and Baishan under Jiangsu's jurisdiction. From the Ministry of the Interior's point of view, the central government's position on the issue had been set, and there was no need to re-investigate the provinces' maritime boundary. However, neither of the provincial governments was willing to accept the earlier agreement. Zhejiang still claimed all the Shengsi Islands, and Jiangsu contended that historical records contained no justification for putting Huanglong Island under Zhejiang's jurisdiction.[64] Eventually, the chairman of the Jiangsu provincial government, Chen Guofu, broke the deadlock by recommending that Zhejiang and Jiangsu send personnel to conduct a survey and agree on the position of their provincial boundary.[65]

In the summer of 1936, Zhao Cisheng met with Jiangsu's Fourth Area administrative supervisor, Ge Tan, to resurvey the boundary. The opposing groups from Zhejiang and Jiangsu who originally sparked the dispute influenced the negotiations from start to finish. At the bequest of Zhao Cisheng, Shanghai's Ningbo Native-Place Association sent the banker Zhang Xiaogeng and other fishing lodge leaders to act as advisors, since they had been most vocal in advocating a change in the Shengsi Islands' administrative jurisdiction. Zhejiang also deputed Chen Tongbai, director of the Zhejiang Provincial Fishery Experiment Station, to assist in the negotiations.[66] In similar fashion, a delegation from Chongming's Nationalist Party branch led by Lu Yanghao met with Ge Tan in Shanghai and expressed opposition to any change in the Shengsi Islands' jurisdiction.[67] Prior to the negotiations, Ge Tan inspected the offices of Shengsi's Fifth Ward, and its leaders urged him not to hand over any of the islands to Zhejiang.

When negotiations between the Zhejiang and Jiangsu delegations got under way at Huanglong Island on July 24, Zhao

Cisheng refused to move forward with the survey unless it re-affirmed the principle of allocating all the Shengsi Islands to Zhejiang. Ge Tan objected that they could not make such a deci-sion. Power to determine Zhejiang and Jiangsu's border rested with the provincial governments, which would reach a decision according to the results of their survey. Unable to get past this disagreement, talks broke down before the survey even began.[68]

With the border question unresolved, the Administrative Yuan held another meeting on September 4 of Zhejiang and Jiangsu, the Ministries of the Interior, Industry, and Finance to reach a final settlement. Hou Chaohai participated as the Minis-try of Industry's representative. Ge Tan and the fishery expert Wang Wentai attended on Jiangsu's behalf. Zhao Cisheng and Chen Tongbai represented Zhejiang. The Ministry of the Inte-rior still wanted to handle the issue according to the 1932 negotia-tions. The Ministries of Finance and Industry advocated keeping the present boundary.[69] After reviewing the various positions, Chiang Kai-shek reversed his earlier decision and ordered the two provinces to maintain the current state of affairs. As a result, the Shengsi Islands stayed under Jiangsu's control.[70] The territorial dispute that resulted from internal divisions within the National-ist regime, the regional loyalties of its leaders, and their divergent economic interests finally came to an end.

In December 1936 Jiangsu sent an investigatory mission to formulate a program for the Shengsi Islands' future development. In addition to Ge Tan and the Chongming Nationalist Party branch's chairman Lu Yanghao, the delegation included promi-nent fishery specialists from Jiangsu and the Nationalist govern-ment, including Hou Chaohai and Wang Wentai. The delega-tion's "administrative group" made plans to tighten bureaucratic control over the Shengsi Islands by implementing the *baojia* mu-tual security system, training militia forces, and heightening offi-cial supervision of fishing organizations. The "economic group" was to reform marketing relations in Shengsi's fisheries. The "fishery group" would carry out investigations of fishery resources and aquaculture experiments, improve marketing and transport, and promote the overall "reconstruction" of the islands' fishing industry.[71]

Zhejiang undertook nearly identical fishery reforms in 1936 with the organization of the Zhejiang Fishery Management Committee (Zhejiang sheng yuye guanli weiyuanhui). The committee's "defense group" took charge of all matters related to fishery protection and the fishery police. The "economic group" handled regulation and reform of the credit system, relief loans, fishery experiments, and improvements in fishing techniques. The "direction group" would compile and inspect *baojia* registers and implement plans to educate and train the fishing population. The direction group would also handle the formation, direction, and management of fishing organizations and the mediation and resolution of fishing disputes.[72] Expert committees were organized to handle specific topics related to fishery development. The "economic committee" included fishery experts such as Zhou Jianyin, as well as fishing lodge directors such as Zhang Shenzhi and Liu Jiting. Other lodge directors were members of the defense group.[73] In April 1937, the Fishery Management Committee initiated annual registration of fishing boats, fish brokers, and fish-processing enterprises in Zhejiang. Enterprises that failed to register could not request the Management Committee's protection or purchase salt used to preserve fish at reduced tax rates. Up to that point, no reliable statistics existed on the number of fishing boats in Zhejiang. Registration was to provide information needed to gauge the progress of the province's fishing industry and promote its development.[74] In addition to boosting fishery production, of course, registration would facilitate taxation of fishing enterprises.

Toward Efficient Exploitation

Beginning in the mid-1930s, China's central government undertook plans to achieve the longstanding goal of more efficient exploitation of the marine environment through rationalized, expert planning. Fishery specialists in the Ministry of Industry took a major step in this direction by preparing for the opening of a centrally run fish market in Shanghai. The Ministry of Industry worked to make the fish market a reality in 1933, when it sent a team of fishery experts led by Hou Chaohai on a fact-finding mission to Japan to gather information on the Japanese government's regulation of the marketing of fish.[75]

Hou stressed that China needed state-controlled fish markets and fishing cooperatives to "achieve the improvement of economic organization in the fishing industry, improve fishermen's livelihoods, and thereby gradually implement control (*tongzhi*) of the fishing industry." At the same time, the Nationalist government had to strengthen fishing organizations to expand productive capacity and create a powerful "rearguard" to support China's relations with foreign countries.[76] Although Hou referred to the necessity of "conducting research on spawning and the protection of fish species," he never explained how these programs for preserving marine resources would be carried out.[77]

Hou Chaohai cited the Nationalist government's financial difficulties as the main cause of its inability to implement fishery reform programs. The government needed to give up on the flawed notion of relying on funds "taken from fisheries and used for fisheries" that was behind its failed attempt to collect fishery reconstruction fees and set them aside specifically to fund fishery administration.[78] Extending centralized control over Shanghai's fish market would generate huge amounts of tax income and give fishery specialists like Hou a reliable source of financial backing. As one foreign observer noted, setting up a fish market in Shanghai promised to give the Nationalist government "an excellent source of revenue."[79] The Chinese state's fiscal objectives again intersected with the drive toward more efficient exploitation.

In 1935 and 1936, the Ministry of Industry used over 1,000,000 *yuan* borrowed from the Ministry of Finance to prepare the new Shanghai Fish Market. The central government's plans were protested by fish brokers in Shanghai, but the Ministry of Industry got around opposition with help from Du Yuesheng and the Shanghai banking magnate Fang Jiaobo. When the Shanghai Fish

Market finally came into existence, eight of its board members came from the business community, and the government selected the other eight. However, the "official" appointees included the bankers Yu Qiaqing and Fang Jiaobo, as well as the fishing lodge director Zhang Shenzhi. Foremost among the business appointees was Du Yuesheng, who was chairman of the board of the Shanghai Fish Market. As an additional concession to Shanghai's fish merchants, the Ministry of Industry allowed business interests to purchase 300,000 *yuan* worth of shares in this centralized market.[80] The Shanghai Fish Market began operations in June 1936 with total capital of 1,200,000 *yuan*, half from the Ministry of Industry and half from business groups in Shanghai.[81]

To overcome seasonal market fluctuations, the Shanghai Fish Market possessed mechanized refrigeration and storage facilities capable of holding up to 1,500 tons of fish for sale during periods of peak demand. With these cold-storage facilities, fishermen would no longer have to sell fish at depressed prices to get rid of their highly perishable catch. The market's ability to smooth over imbalances in supply and demand promised to stabilize fish prices and increase production.[82] The Fish Market's regulations required all fish sold in Shanghai to pass through the state-run marketing facility. Only brokers affiliated with the market could purchase the catch from fishing boats. By controlling the sale of fish in China's largest consumption center, the Shanghai Fish Market intended, according to one foreign commentator, to "centralize, if not monopolize, the fishing industry."[83]

The Shanghai Fish Market charged a commission of 7 percent on the value of all fish products sold, and the government prohibited any fees except for ones authorized in the market's regulations. The Fish Market claimed 4 percent, and the remaining 3 percent went to the brokers who took part in the transaction. The merchant middlemen also took a handling fee of 1.55 percent.[84] Since all imported fish products had to pass through the centralized Shanghai Fish Market as well, the Nationalist government secured additional revenues by ensuring that Japanese boats carrying fish products paid the required customs duties.[85]

The Nationalist government's fishery experts moved even closer to the elusive goal of rationalizing use of marine resources by reforming the credit relations that gave fishing enterprises their capital. Responding to appeals from fishing lodges in the Zhoushan Archipelago, whose fortunes took a major downturn after the global depression fully impacted China's economy in 1934, the Ministry of Industry brought together a consortium of financial institutions in Shanghai in the fall of 1936 to form the Fishery Bank (Yuye yintuan) to issue loans to fishing enterprises. The banker Kui Yanfang, whose investments in fish brokerages had gained him a position as a director in the Shanghai Fish Market, drafted the Fishery Bank's regulations.[86] Beginning in October 1936, the Fishery Bank made loans totaling 120,000 *yuan* to fishing cooperatives in Yin county and brokers affiliated with the Shanghai Fish Market.[87] This financial relief from the Nationalist government stimulated a recovery in fishery output and sustained ecological pressures on fish stocks.

Zhoushan's Fisheries During the Sino-Japanese War

The onset of war between China and Japan in 1937 had severe consequences for Zhoushan's fishing industry. In the summer of that year, the Japanese imposed a blockade on coastal shipping, making it impossible for Chinese fishing boats to enter the port at Shanghai.[88] Japanese forces dominated the waters around the Zhoushan Archipelago, occupying the Shengsi Islands as a base for aerial assaults on Shanghai, Nanjing, and Hangzhou. The Japanese brought interprovincial squabbling over the Shengsi Islands to an end by occupying most of the Zhoushan Archipelago in 1939.[89]

After the Japanese invasion, official regulation of Shanghai's fish market assumed a new form. In November 1938, Japanese authorities set up the Central China Marine Products Company (Ch. Huazhong shuichan gufen youxian gongsi; J. Kachū suisan kabushiki kaisha) to monopolize fish wholesaling in Shanghai and license Japanese trawlers operating out of the port. The Japanese justified their monopoly by saying that Chinese fishermen's inefficient methods for transporting and distributing fish prevented their rational use and resulted in the "desecration" of natural resources. State regulation—under Japanese auspices, of course—was therefore a necessity.[90] During the war, revenues obtained by regulating the sale of fish in Shanghai went to the Japanese navy, which strictly enforced controls over marketing.[91] Nevertheless, direct links existed between the Central China Marine Products Company and the Nationalist government's Shanghai Fish Market. The Chinese fishery expert Zhang Zhuzun worked with the Japanese-controlled marketing organ as head of the Department of Fisheries and Animal Husbandry in the collaborationist Wang Jingwei regime's Ministry of Industry.[92]

Following the opening of the Central China Marine Products Company, Japanese authorities loosened their shipping restrictions and permitted Chinese boats to land fish in Shanghai if they sold their catch through the new marketing organ.[93] Mechanized Japanese vessels and Chinese fishing boats that obtained licenses from the Japanese found a ready market in Shanghai, where the influx of refugees to the city's International Settlement after the outbreak of war had led to an upsurge in demand. As a result, fish consumption in Shanghai, most of which was still brought to port by Chinese boats that fished in the waters of the Zhoushan Archipelago, exceeded prewar levels until 1941.[94]

To compete with the Japanese for control of the fish supply, brokers in Shanghai who participated in the Nationalist government's Shanghai Fish Market teamed with foreign business interests in June 1938 to form the Sino-French Fishery Company. This joint venture handled the sale of fish in Shanghai's French Concession, which had not yet fallen to the Japanese.[95] To gain access to a supply of fish, the Sino-French Fishery Company sent personnel to extend 20,000 *yuan* in loans to fishing enterprises in Zhoushan and confer with fishing industry leaders in Shenjiamen, such as Zhang Xiaogeng, about transporting fish to Shanghai.[96] The Sino-French Fishery Company also relied on its European

partners to protect shipments of fish against Japanese interference.[97]

Prior to 1941 much of coastal Zhejiang was still under the control of the Nationalist regime. For fishermen in Nationalist-held
regions, the Japanese naval presence made their occupation even
more precarious than usual. As a 1939 investigation of economic
conditions in Fenghua county observed:

Because of the influence of the war, all coastal fishermen were at first
afraid to go out to sea. Later, some of them had no choice but to fish for
their livelihood. They are often held under duress by enemy warships,
and sometimes they are forced to transport their catch to Shanghai and
sell it to the fish market that the enemy has set up. Therefore, the economic situation is increasingly downcast compared to previous years.[98]

Boats that went on fishing during the war were also threatened by
pirate bands that ran rampant off the coast of Zhejiang, seizing
fishermen's money and supplies.[99] In Taizhou's Wenling county,
wartime disruptions caused many fishing boats to stop making
their seasonal migration to the Zhoushan islands for the winter
hairtail season.[100] Shanghai's Dinghai Native Place Association
appealed to Nationalist authorities in Ningbo on behalf of fish
merchants hoping to obtain steamships to transport their catch
from Dinghai to Shanghai for sale.[101] Fishing enterprises throughout coastal Zhejiang faced similar difficulties during the early
years of the war, and fishery production seriously declined in Nationalist-held regions.[102]

Whenever they encountered Japanese military forces, ice boats
had to hide the permits issued by the Nationalists or simply
throw them into the water in a panic. Otherwise, the Japanese
would search, confiscate, and even destroy their cargo. Ice boats
were not set free unless they returned to Shanghai to sell their
fish through the Japanese-controlled marketing organ. Once ice
boats made it to the port of Zhenhai, they had to pass through
military and customs inspections and pay the local fishing association a fee of 1 to 5 *yuan*. Only then could they go to Ningbo
and sell their catch to fish brokers. After the Nationalists imposed a blockade on Zhenhai, fishing boats had to stay for up to a

month after selling their fish unless they found a way to leave the port secretly. In April 1939, Japanese bombing raids demolished practically all the brokerages in Ningbo's fish market. Supply did not meet demand, which resulted in fish prices up to five times higher than normal. High prices motivated ten or so fish brokers to set up sheds and resume business at their original locations.

Once Dinghai had fallen to the Japanese, fishing boats had to obtain passage permits (*tongxingzheng*) from the occupiers. To obtain permits, fishing boats were required to have a fish broker in Shenjiamen to serve as their guarantor. Fishing boats faced strict inspections, and those that did not possess Japanese-issued permits could have their property confiscated or meet with other difficulties. Chinese collaborators and Japanese *rōnin* in charge of the Fishery Office (Yuyebu) set up by Shenjiamen's Self-Government Association (Zizhihui) also collected fees from fishing boats. To make matters worse, the Japanese forcibly recruited many able-bodied fishermen as conscripted laborers. Faced with this harassment, "many fishing boats have fled to other places, and boats from other fishing groups, because of various restrictions, cannot go all the way to Dinghai to fish. Therefore, fish brokers in Dinghai's port of Shenjiamen will also decline and go out of business."[103] In August 1938 a report in *Dinghai min bao* estimated that, since the start of hostilities with Japan in 1937, Zhoushan's fishing industry had suffered losses of 2,368,000 *yuan*.[104]

Fishery experts in Zhejiang and the Nationalist government, such as Hou Chaohai and Jin Zhao, tried to maintain fishery production by setting up cooperatives to provide relief loans to fishing enterprises.[105] Zhejiang actively implemented these policies between 1940 and 1942, offering over 200,000 *yuan* in loans each year to fishing cooperatives in areas under Nationalist control. But after most of Zhejiang fell to the Japanese during the Ningbo-Shaoxing Campaign of 1941 and the East China Campaign of 1942, it was impossible for the province to continue these relief programs, and it stopped issuing fishery loans entirely during 1943 and 1944.[106]

The military conflict between China and Japan in the 1940s was a serious blow to fishing activities in the Zhoushan region. Of the roughly 26,000 boats that had fished in Zhejiang prior to 1937, approximately 15,000 were destroyed during the war.[107] According to one estimate, the prolonged military confrontation with Japan and the Chinese Civil War that followed caused catches by Chinese fishing boats in Zhoushan to fall from their prewar level of 93,000 tons in 1936 to a mere 12,000 tons in 1947.[108]

The Japanese fishing industry fared little better during the 1940s. Virtually all fishing by Japan's mechanized fleet came to an end in 1942 with the onset of the Pacific War. A report from the Central China Marine Products Company's 1945 shareholders meeting explained that fish supplies had fallen dramatically due to the difficulty of procuring materials and the hazards that the war presented for transporting the catch.[109] The Japanese navy drafted many fishing vessels for military service, and the demands of total war did not leave enough raw materials for those that could still engage in business.[110] Even though it brought economic disaster for many fishing enterprises during the 1940s, the precipitous wartime decline of the Japanese and Chinese fishing industries gave fish stocks a temporary respite from human demands.

Conclusion

During the mid-1930s, local governments in Jiangsu levied taxes on fishing boats from Zhejiang, and Zhejiang taxed boats that fished in Jiangsu's provincial waters. This complicated situation would not have existed if shifts in the position of fishing grounds had not forced boats from Zhejiang to move into waters within Jiangsu's borders in search of productive stocks. In this way, the unpredictability of the marine environment set the stage for political confrontation. However, the immediate cause of the dispute between Jiangsu and Zhejiang was the Chinese state's effort to increase extraction of revenues at all levels of society. As a result of these fiscal demands, provincial and local governments in Zhejiang and Jiangsu competed as vigorously for economic returns obtained from the marine environment as opposing regional fishing groups. These bureaucratic objectives made it virtually impossible to reach viable agreements coordinating the use of Zhoushan's fisheries.

Despite their differences, all the competing interests that made up the Chinese state favored developmental policies that expanded their tax base by promoting efficient exploitation of the natural environment. The government could not tax a boat that did not cast its nets or a fish that stayed in the water. For this reason, Nationalist officials had little incentive to limit catches. Regulations that required producers to forgo exploitation of natural resources only decreased tax receipts. Of course, the total exhaustion of fish stocks could have eliminated this source of revenues altogether, but official desire for fiscal income discouraged state actors from placing much value on these long-term considerations. The desire to convert natural resources into revenues exerted just as strong an influence with the Nationalist regime as it did under the Japanese occupation. Ultimately, only the devastation caused by total war alleviated demands on the marine environment. War had devastating consequences for China's natural landscape, but it was an ecological respite for fish populations.

Continuities and Discontinuities

Throughout the twentieth century, Chinese fishery experts advocated the use of modern science and technical expertise to bring about a more rational and efficient exploitation of the marine environment. Before 1949, fishery experts fell short of implementing their plans to remake relationships between society and nature in the Zhoushan Archipelago. Nevertheless, their developmental vision exerted a direct influence over fishery policies carried out under the PRC in the 1950s. Judith Shapiro has shown that the "polarizing, adversarial" conception of the natural environment that characterized Maoist ideology lay behind much of the ecological devastation experienced in China after 1949.[1] But in Zhoushan's fisheries, the PRC state's policies toward the marine environment grew directly out of blueprints drawn much earlier in the twentieth century. The goal of rationalized, efficient exploitation of natural resources shaped the Chinese state's developmental policies before and after 1949. These developmentalist objectives coincided with the twentieth-century project of modernity that prevailed throughout the world under a wide variety of political and economic conditions.

Once it took control of the Zhoushan islands in 1950, the communist government made substantial loans to fishing enterprises to encourage prompt recovery of fishery production. Between 1950 and 1957, this credit relief amounted to 12,439,000 *yuan*.[2] Government-organized fishing cooperatives took over provision of gear and supplies to further promote increased output.[3] The PRC also realized Republican-period plans when it set up state-run marketing facilities in Shenjiamen and ports in the Shengsi Islands to control pricing and distribution of the catch. Only fish brokers who registered with local authorities and reported information on all their transactions to the marketing organs were permitted to stay in business.[4]

By encouraging the addition of outboard motors to sail-powered fishing boats—a technological improvement initially suggested by Chinese fishery experts in the 1930s—the PRC government facilitated more intensive harvesting of Zhoushan's fish stocks.[5] Motorized junks cost only one-tenth as much as steam trawlers, but were two to three times more productive than sail-powered fishing boats.[6] To raise capital for motorized boats, fishing cooperatives launched campaigns urging female members of fishing households to step up production of rope and fishing nets as a side-employment.[7] Between 1956 and 1963 the number of motorized boats in Zhoushan jumped from 76 to 1,200, and by 1963 they accounted for 40 percent of the total catch.[8] Technological innovations that boosted production by expanding into more distant seas and targeting more lucrative fishing grounds persisted into the PRC period from late imperial times, when migrant fishermen brought more effective boats and fishing gear from their native regions to the Zhoushan islands.

What is more, the PRC fulfilled developmental goals that dated to Sun Yat-sen's *Industrial Plan* by constructing a modern fishing harbor on Shengshan Island in the early 1960s.[9] All these fishery development projects grew out of the drive to rationalize exploitation of the natural environment that began in the Republican period. This striking continuity in the Chinese state's fishery policies across the 1949 divide is not at all difficult to explain, since the same people were responsible for formulating them. Fishery specialists like Li Shixiang, Hou Chaohai, and Jin Zhao who staffed fishery offices under the Nationalist regime held influential leadership positions in PRC government agencies well into the 1950s.[10] The early years of the PRC witnessed the culmination of longstanding efforts by these fishery experts to remake human interactions with the marine environment.

The vital difference before and after 1949 was that the PRC, unlike its Nationalist predecessor, had the capacity to put these measures into effect. The communist party-state deputed teams of cadres to ensure its initiatives were enacted at the local level. Work committees registered and gathered financial information on fish brokers in the Zhoushan Archipelago and carried out household registration.[11] The Chinese Communist Party's local committee in Zhoushan carried out a "fishery reform" (*yugai*) modeled on China's land reform of the early 1950s. But to speed the recovery and development of Zhoushan's fishing industry, the fishery reform followed an extremely lenient class line. Local cadres chose to unite with "fishing capitalists" (*yuye zibenjia*) and directed class struggles only against "fishing bullies" (*yu ba*) found guilty of "political" as well as economic oppression.[12]

To prevent the jurisdictional disputes that plagued the Nationalist government during the 1930s from resurfacing, the PRC unified administration of the Zhoushan Archipelago in March 1953 by placing the Shengsi Islands under the Zhoushan Prefectural Commissioners Office (Zhoushan zhuanshu).[13] The Prefectural Commissioners Office and counties along the coast of Zhejiang set up fishery production direction offices (*yuye shengchan zhihuibu*) to supervise fishing activities, set up mutual-aid cooperatives, oversee transport and marketing, and maintain public order. When fishing season arrived, personnel from these offices went to fishing grounds to direct production and resolve problems. Boats still fished in fleets defined by geographical origin, but local cadres selected activists (*jiji fenzi*) from among fishing communities to lead their compatriots and make sure they followed official directives.[14] By expanding its reach in local society and tightening the mechanisms of control, the PRC state enhanced its ability to carry out developmental initiatives.

With the implementation of these fishery reforms, catches quickly returned to pre-1937 levels. Nature assisted as well, since high levels of rainfall and strong monsoon winds during the 1950s made environmental conditions favorable to fishery production.[15] Fishery output in the Zhoushan region reached 83,000 tons in 1952 and continued to increase throughout the decade.[16] The easing of human pressures on the marine environment during the military conflicts of the 1940s had also given fish stocks a chance to replenish, and seasoned fish brokers in Zhoushan commented that they had never seen so many fish.[17] Favorable climatic conditions and recovery of fish populations brought greater catches for fishing enterprises, at least for a time. The PRC benefited from increased output by collecting a 5 percent tax on fish sold at state-run marketing organs, which it later raised to 8 percent in the spring of 1956.[18] Local cadres in the islands supervised fish merchants and made sure they did not evade taxes.[19] As in the Republican period, the Chinese state's desire to gain tax revenues was an important motivation to make exploitation of natural resources more efficient. With developmental programs increasing catches and bolstering fiscal income, state agencies held to their belief that scientific and technical expertise would make natural resources endlessly productive. Fishery experts still put their faith in prohibitions on types of gear that damaged spawning activities, promotion of more efficient technologies, aquaculture, research on the distribution and fluctuations of fish stocks, and the opening up of deep-sea fishing grounds to perpetually expand production.[20]

On the other hand, the increasing pressures on Zhoushan's fisheries during the 1950s generated intense competition for resources, giving rise to frequent disputes. Cadres deputed to fishing grounds mediated these conflicts and kept production going smoothly. In contrast to the Nationalist period, more effective coordination and communication among various levels of administration facilitated resolution of resource conflicts. Yet state guidance did not always eliminate tensions. In 1954, for instance, Zhejiang province gave in to calls from the Taizhou Prefectural Commissioners Office and delayed a proposed ban on cuttlefish cages so that Taizhou fishermen who had not yet changed trades could participate in production. The Zhoushan Prefectural Commissioners Office refused to follow this decision, drawing protests from local governments in southern Zhejiang.[21] Under the PRC, divergent regional and bureaucratic interests could still prevent the implementation of higher-level directives.

Nor did the establishment of New China in 1949 eliminate international pressures from a rebuilt and even larger Japanese fleet, which returned to fishing grounds off the coast of China in the 1950s.[22] Fishing disputes between China and Japan flared throughout the decade.[23] To make matters worse, the small fleet of mechanized trawlers operated by state-owned Chinese fishing enterprises often ignored official prohibitions and entered coastal waters, damaging traditional fishing vessels. The PRC's fishery management agencies expressed concern that the internal disputes that resulted from these incursions would "influence our country's international credibility, which is detrimental to the fishery struggle against Japan."[24] Domestic and international pressures on Zhoushan's marine resources cut across the 1949 divide, but in the 1950s and 1960s the level of exploitation had grown to unprecedented levels.

Before long, expanding human demands on the marine environment had an impact on Zhoushan's most economically important fish species. As early as 1963, reports noted that the size of large and small yellow croakers landed by fishermen in the Zhoushan region had decreased substantially since the 1950s.[25] However, many PRC officials firmly maintained that China's fish stocks were still abundant. The only problem, in their view, was that fishermen did not exploit these resources rationally. Much as in the Republican period, PRC fishery management agencies pointed to wasteful, inefficient fishing techniques and Chinese fishermen's inability to open up new, unexploited fishing grounds as the cause of declining catches. As a Zhejiang fishery management office stated in 1968, "The best method for protecting resources and solving contradictions is to actively open up new resources and fishing grounds."[26] Central government regulations called for research to protect immature fish in spawning grounds, but restrictions were never enforced. Increasing production always held higher priority than preserving fish stocks.[27]

By the 1970s, populations of large and small yellow croaker, cuttlefish, and hairtail showed signs of severe depletion. Marine pollution caused by rapid industrial development in coastal areas further contributed to this decline. Alongside these anthropogenic factors, lower precipitation and weaker monsoons during the 1970s made fishing grounds less productive.[28] By the early 1980s, the annual spawning runs of yellow croaker and cuttlefish that attracted fishers to waters off the Zhoushan Archipelago during the Qing dynasty had ceased, and hairtail were extremely scarce as well. During the 1970s Zhoushan's main commercial species—small yellow croaker, large yellow croaker, hairtail, and cuttlefish—made up 60–70 percent of the total catch. In the 1990s these species accounted for only 20 percent of total output. By the end of the twentieth century, large and small yellow croaker and cuttlefish stocks had collapsed, and hairtail were in serious jeopardy. For a short while, declines in yellow croaker and hairtail populations gave crab, shrimp, and smaller fish that croakers and hairtail prey on a chance to increase. But commercial exploitation quickly started to deplete these prey species as well, which hurt the recovery of fish that rely on them for food.[29]

Since the Republican period, Chinese fishery experts had expected modern science and technology to compensate for declining yields that result from heightened exploitation of finite common-pool resources. As far back as the 1920s and 1930s, fishery experts perceived diminishing catches in Zhoushan's coastal fishing grounds. But they never questioned the assumption that modern science and technology would manipulate the ocean's resources for maximum productivity. Under certain circumstances, as seen in the cuttlefish feud, this emphasis on efficient use led fishery experts to advocate restrictions on some types of fishing. But the purpose of making exploitation of resources more rational and less wasteful was to achieve national economic development, increase production, and maximize fiscal returns. Under

the PRC, the intensified extraction of marine resources undertaken in pursuit of these goals led to the collapse of the Zhoushan
Archipelago's most important fish species. Ultimately, belief in
the power of modern science hastened rather than averted environmental degradation.

Despite their claim to detached objectivity and scientific cer-
tainty, Chinese fishery experts never espoused a single, clear-cut
solution to the problem of regulating marine fisheries. The formation of fishery management policy was always inherently political and influenced by economic interests. During the 1930s,
regional coalitions of fishermen who used different types of gear
exerted constant pressure on state agencies to shift policy in their
favor. When fishery experts tried to gather information on fish
populations to settle disputes between feuding native-place
groups, research gave way to further bureaucratic infighting.
With an array of state actors struggling to garner support for
their favored policies, resolution of fishing controversies degener-
ated into struggles for budgetary appropriations. Funding priorities surfaced again during border disputes between Zhejiang and
Jiangsu during the mid-1930s, in which each province vied for a
greater share of the profits extracted from Zhoushan's marine environment.

When handling environmental issues, Chinese officials—not
unlike bureaucrats elsewhere in the world—were often guided by
a need to maintain and expand jurisdictional turf and budgets.[30]
Fishery management in modern China, echoing Tim Smith's
characterization of fishery science in the United States and
Europe during the late nineteenth and early twentieth century,
was "dictated by transitory economic and political forces."[31]
Boosting production and revenues in the short-term generated far
less social and political resistance than restrictions that advan-
taged some at the expense of others. Unable to agree on the appropriate way to distribute access to common-pool resources,
Chinese fishery experts devoted their energies to less controversial programs that heightened exploitation of marine resources
with little regard for their preservation.

Local forms of regulation enacted by fishing communities in the Zhoushan Archipelago during the Qing and Republican period were based on a different understanding of the marine environment, but one that was far from ecologically benign. To fishermen, the ocean was a world fraught with uncertainty and risk. This insecurity derived from the dangers and risks of life at sea, as well as the problems that characterize common-pool resources. Local religion protected fishermen's welfare and security in this chaotic, unpredictable, and hazardous environment. By enacting rules to coordinate the use of fishing grounds, native-place coalitions prevented the loss of economic returns to violent conflict. Without such regulations, pervasive uncertainty and conflict would have made profitable harvesting of fish all but impossible. These interpenetrating religious beliefs and social institutions, to borrow Robert Weller's phrase, were "localist and human-centered."[32] Temple religion and social institutions focused on the welfare of particularistic human communities. Thus, it would be a mistake to idealize these environmental understandings. Neither local religion nor customary regulation expressed a concern with preservation of the natural environment for its own sake.

Throughout China's late imperial and modern periods, human interactions in the marine environment entailed a vigorous pursuit of profit. Commercial integration and the development of financial networks made it possible for small-scale producers to exploit fish stocks more and more intensively, putting greater demands on limited resources. Improved fishing technologies undergirded this intensified exploitation, allowing expansion into more distant fishing grounds as the productivity of inshore stocks declined. As the value of fish stocks increased, native-place groups organized to secure benefits they derived from common-pool resources against outsiders. In Zhoushan, local militarization and environmental change went hand in hand. In response to heightening competition during the final years of the nineteenth century, regional coalitions formed militias to uphold proprietary claims to fishing grounds.[33] Yet from the late Qing into the Republican period, mediation by the elite leaders of native-place groups usually diffused conflicts over resources and kept violence at a minimum. Local officials sometimes had to intervene when internecine competition escalated into armed feuding. But officials depended on fishing lodge leaders to broker settlements and maintain order. At other times, local elites resolved fishing disputes independently and received official approval for settlements after the fact.

Social institutions that native-place groups in Zhoushan crafted to cope with heightened competition for a limited body of resources resembled what Arthur McEvoy calls the "ad hoc, quasi-legal systems" of regulation that ethnic coalitions of immigrant fishers in California relied on to "insulate themselves from the market forces that drive unorganized fishers collectively to ruin both their resources and their livelihoods."[34] However, an interpretive framework that draws a sharp distinction between "traditional" forms of regulation and the forces of commercialization cannot be applied to the history of Zhoushan's fisheries. The patterns of regulation that existed in Zhoushan during the late Qing and Republican period did nothing to shield marine resources from market pressures. These institutions safeguarded profits extracted from the marine environment by limiting violent conflict but did not alleviate the ecological impact of economic integration. Commercial networks that transformed fish into marketable commodities depended on these social institutions to function. Doing business with compatriots from one's own native-place group also made it less costly to evaluate the reliability and trustworthiness of trading partners, get access to stable sources of supply, enforce compliance with agreements, and obtain information about market conditions.[35] The objective was profitable exploitation of fish stocks, not their preservation. The post-1949 period saw an intensification of patterns of ecological degradation, but not a change in kind. Rules that limited competition and diffused violent conflict prevented costly disruptions in fishery production, keeping extraction of resources and their transformation into commodities going smoothly.

Of course, profits gained through commercial exploitation of Zhoushan's fisheries did not accrue to everyone equally. Social institutions mirrored relations of stratification and power within the native-place community. A disproportionate share of the benefits went to elites who acted as fishing lodge leaders. By coordinating use of fishing grounds and mediating disputes, the elite leaders of different native-place coalitions gained a steadier and more reliable flow of profits from Zhoushan's fishing industry. These benefits gave Chinese elites a strong incentive to enforce regulations governing fishing activities. Their power and influence was also necessary to secure cooperation and compliance with these rules. Unofficial regulation reinforced relations of power, ensuring that sizable economic returns accrued to the commercial and financial elites who came to dominate the region's fishing industry in the Republican period.

By the early twentieth century, the growth of fishery production in Zhoushan appears to have already had an impact on the marine ecosystem. Environmental conditions led to frequent fluctuations, but signs of human overexploitation were undeniable. Fishermen caught larger numbers of small, immature fish, catches from intensely exploited inshore fishing grounds declined, and boats ventured into distant waters in pursuit of more productive resources. Human-induced ecological changes, in turn, had important social repercussions. As fish stocks grew scarce relative to human demand, heightened competition increased the potential for conflict. Tensions between regional fishing groups grew more entrenched with every dispute, making settlements more difficult to achieve. By the early 1930s, local regulation could no longer prevent violent conflict over finite natural resources.

With opposing native-place coalitions struggling over diminished resources, the Zhoushan Archipelago started to resemble other regions of China that had been densely settled for longer periods of time. Harry Lamley, for instance, characterizes the lineage feuding that plagued mainland Fujian and Guangdong as "an extreme form of competition for resources in a relatively overpopulated region."[36] With population expanding and resources in short supply, no customary rules or obligations could settle violent conflicts and reimpose order.[37] Sow-Theong Leong found similar links between ecological change and social conflict in southeast China, where Hakkas and other Chinese communities peacefully coexisted during economic upturns when uncultivated land existed for settlement. Whenever economic opportunities declined and land grew scarce, ethnic conflicts quickly flared.[38] Under conditions of severe ecological scarcity and degradation, as Elizabeth Perry discerned in North China, people "found it necessary to seize or defend diminished resources by violent means."[39]

Regulations enacted by native-place groups in Zhoushan could minimize local disputes over fishing grounds, but local arrangements proved incapable of coping with larger-scale ecological problems. Internationally, the Japanese fishing fleet's exhaustion of highly valued fish species in the East China Sea during the mid-1920s drew its mechanized trawlers into fishing grounds off the Zhoushan Archipelago. Domestically, long-term processes of demographic expansion in peripheral regions of southern Zhejiang outstripped available land and contributed to the breakdown of water-control systems. In the wake of serious flooding, households from Wenzhou and Taizhou turned to fishing with cages as a lucrative source of profits. Existing social institutions had no way to accommodate competitors from these coastal peripheries and their new technologies. At the same time, international law and power politics made it impossible for the Chinese government to exclude Japan's mechanized fleet from offshore waters frequented by domestic fishing boats. Foreign and domestic competitors reduced returns that other users derived from fish stocks, generating fierce opposition. Since the late Qing, Zhoushan's fishing industry had placed considerable pressures on the marine environment. Mechanized trawlers and cuttlefish cages accelerated this ecological change.

Along with intense foreign and domestic competition, changing relations between regional associations and China's larger political system in the early 1930s helped to break down social institutions coordinating the use of Zhoushan's fishing grounds. Since the Qing, native-place coalitions in Zhoushan had received official sanction and approval for their regulations coordinating the use of fishing grounds. Local governments supported and benefited from these arrangements, since regional fishing groups and their elite leaders helped maintain local order and collect tax revenues. Unable to deal with Japanese trawlers and cage fishers through existing rules governing the use of resources, fishing groups called on the Nationalist government to exclude outside competitors. At this point, only vigorous official initiatives would have been able to at least slow the exploitation of Zhoushan's fisheries. But unlike the glory days of Qing statecraft, enlightened official action was a rarity in local politics during the Republican period. Instead, state interventions and the interests that motivated them only ended up aggravating conflicts over natural resources.

The modern Chinese state's expanded presence in local society during the early twentieth century, which Prasenjit Duara terms "state involution," multiplied the number of groups claiming the power to administer and tax Zhoushan's fishing industry.[40] In the Republican period, the presence of overlapping state agencies at the local level significantly increased the potential for violence and conflict. During the cuttlefish feud, the leaders of competing native-place groups used the payment of fees to persuade state actors to support their claims. Later in the 1930s, when Zhejiang and Jiangsu vied for revenues exacted from fishing enterprises, regional groups that local governments relied on to collect fees were also embroiled in the administrative disputes. With the modern Chinese state's ever-increasing demand for revenue, collaboration between native-place groups and state actors intensified contests for control of scarce resources.

As Mark Elvin has observed, the state's interest in "the creation and appropriation of a growing and dependable supply of tax-resources, soldiers, and conscripted labour" has dominated official decision-making regarding the environment over the course of Chinese history.[41] China's modern state expanded these objectives, seeking to control and manipulate nature for maximum production and revenues. Processes of state building in modern China created an unprecedented demand for funding that amplified the fiscally driven exploitation of nature. The rational application of modern science and technology, it was assumed, would overcome every obstacle to increasing economic production and maximizing revenues. As a result, state agencies favored developmental programs that realized immediate returns by intensifying exploitation but gave little thought to the possibility that this growth might not last. Throughout the twentieth century, as J. R. McNeill observes, governments sought to "maximize their current formidability and wealth" in the short-term even if they risked "sacrificing ecological buffers and tomorrow's resilience." Pursuing these objectives, governments all over the world moved "to make the most of resources, make nature perform to the utmost, and hope for the best."[42] China was no exception.

Zhoushan's marine environment fared poorly under local forms of regulation. It fared worse under centralized state management. Demographic pressure, commercial integration, increased mobility, technological innovations that facilitated ever-more intensive extraction, and the presence of state agencies seeking to extract greater tax revenues generated intensified exploitation of fish stocks. Commercial development was the thread that tied these variables together. Population grew in response to opportunities presented by commerce, and people throughout China migrated far and wide to take advantage of these economic opportunities. The modern Chinese state, in all its various incarnations, obtained key sources of revenue from taxes on trade. Even the PRC's economic reforms of the 1950s built on commercial networks that had taken shape during the late nineteenth and early twentieth century.

Today, the Chinese government finally recognizes the severe overexploitation of its fisheries. Official directives seek to protect threatened fish populations and promote sustainable development by reducing fishing effort through restrictions on numbers of boats and gear. Instead of capture fisheries, the industry now looks to aquaculture, which contributes to marine pollution in the form of fertilizers and uneaten fish food, to increase production. Government campaigns praise fishermen who have changed occupations and found jobs in other economic sectors. However, many obstacles still stand in the way of these initiatives' success. Drawn by the upsurge in fish prices since China's economic reforms of the 1980s, residents of coastal and inland agricultural regions have invested in boats and started to fish inshore waters in large numbers. These new entrants confound official conservation efforts and place even greater demands on Zhoushan's strained fish stocks. Local officials engaged in a constant struggle for financial resources commonly engage in "protectionism" and permit these illegal activities.[43]

Even with the tremendous changes apparent in contemporary China, the roots of at least one of its widespread environmental problems display a troubling persistence. As in the Republican period, the contemporary Chinese state seeks to solve environmental problems by relying solely on technical expertise and official regulations. These top-down interventions overlook the creative, adaptive strategies that people in China have developed to deal with ongoing environmental changes. Needless to say, these ecological strategies have had serious limitations. But the Chinese state's environmental policies might prove more effective if they recognize existing social practices and try to point them in more sustainable directions, while also helping to devise realistic alternatives to ecologically damaging activities.

Reference Matter

Chinese Characters

Entries are alphabetized letter by letter, ignoring word and syllable breaks with the exception of personal names, which are ordered first by the surname and then alphabetically by the given name.

Chunhu 蒓湖
Cixi 慈溪

dabuchuan 大捕船
da duichuan 大對船
da huangyu 大黃魚
Dai Yongtang 戴雍唐
Daishan 岱山
daiyu 帶魚
Dalian 大連
danmin 蛋民
Dechang General Store 德昌
　雜貨店
dengji tongxunchu 登記通訊處
diaoyu 鯛魚
Dinghai 定海
Dinghai min bao 定海民報
Dinghai yuye chuanxisuo 定海
　漁業傳習所
Dinghai Zhou bao 定海舟報
Ding Shen dui yuye gonghui
　定沈對漁業公會
Ding Shen yuzhan gonghui
　定沈魚棧公會
diqi shiran 地氣使然
di wu tequ xingzheng ducha
　zhuanyuan 第五特區行政
　督查專員
Dongmen 東門
Dongqian Lake 東錢湖
Dongshajiao 東沙角
dongshi 董事
Dongyang 東陽
Du Yuesheng 杜月笙
Dui yuye gonghui 對漁業公會
Dunhe Lodge 敦和公所

Fang Jiaobo 方椒伯
fatuan 法團

Fenghua 奉化
Fenghua huiguan 奉化會館
Fengnan Lodge 豐南公所
feng wu keshui 封物課稅
fenhui 分會
Fuhai 福海
Fuzhou 福州

Gaoting 高亭
Ge Liquan 葛醴泉
Ge Tan 葛譚
gedi renmin liti jingying 各地
　人民立體經營
gongan fenju 公安分局
gong chuan 公川
Gonghai yuye jiangli tiaoli 公海
　漁業獎勵條例
gonghui 公會
Gongmen 宮門
gongsuo 公所
gongsuo xiansheng 公所先生
gongyi juan 公益捐
Gouqi 枸杞
Guangfuhui 光復會

Haiguo tuzhi 海國圖志
Haijiao 海礁
Hai Longwang 海龍王
Haimen 海門
haiquan 海權
Haizhou 海州
Haizhou yuye chuanxisuo 海州
　漁業傳習所
Han Yougang 韓有剛
hanghu yushou 行户漁首
Hangzhou 杭州
He Huiyu 何恢禹
He Yushu 何玉書
He Zhizhen 何之貞

Hengjie 橫街
Hengshun Lodge 恆順公所
Hou Chaohai 侯朝海
Hu Baoquan 胡寶泉
Hua Ziqing 華子清
Huang Jinrong 黃金榮
Huanglong 黃朧
Huangyan 黃岩
Huaniaoshan 花鳥山
Huazhong shuichan gufen
　youxian gongsi 華中水產
　股分有限公司
Huiguan 會館

Jiang Beiming 江北冥
Jiang Jieshi 蔣介石
Jiangshan 姜山
Jiangsu baowei yuye ju 江蘇
　保衛漁業局
Jiangsu shengli yuye shiyan-
　chang 江蘇省立漁業試驗場
Jiangsu shuichan xuexiao 江蘇
　水產學校
Jiang-Zhe qu yuye gaijin wei-
　yuanhui 江浙區漁業改進
　委員會
Jiang-Zhe yuhui 江浙漁會
Jiang-Zhe yuye gongsi 江浙
　漁業公司
Jiang-Zhe yuye guanli ju 江浙
　漁業管理局
Jiang-Zhe yuye shiwu ju 江浙
　漁業事務局
Jiaotong yinhang 交通銀行
jieyan 戒嚴
jiji fenzi 積極分子
Jin Zhao 金炤
Jinghe Lodge 靖和公所
ju 橘

jueji yu haiyang 絕跡於海洋

Kachū suisan kabushiki kaisha
　華中水產株式會社
Kanmen 坎門
ke bang 客幫
Kong Xiangxi 孔祥熙
Kui Yanfang 簣延芳

Langgang 浪崗
Lao yushang gongsuo 老魚商
　公所
Li Shixiang 李士襄
lihai chongtu 利害衝突
lijin 釐金
ling 靈
Ling Pengcheng 陵鵬程
Linhai 臨海
Linhai xian yuhui 臨海縣漁會
linshi yingyeshui 臨時營業稅
liquan waiyi 利權外溢
Liu Boyu 劉伯瑜
Liu Huafang 劉華芳
Liu Jiting 劉寄亭
Liu Tongjiao 劉同蛟
Liu Xiaosi 劉孝思
liu si hang 六四行
liuwang chuan 流網船
Longdaohui 攏刀會
Lou Yuren 樓谷人
Lu Yanghao 陸養浩
Lühua 綠華
Lü Sheng gongsuo 旅嵊公所
Lüsi 呂泗
Lü Yong Taizhou gongsuo 旅甬
　台州公所

Majishan 馬跡山
Mazu 媽祖

meizi 梅子
mengya 萌芽
Miaozihu 廟子湖
moyu 墨魚

Nanding Lodge 南定公所
Nanpu Lodge 南蒲公所
Ningbo 寧波
Ningbo lü Hu tongxianghui
　寧波旅滬同鄉會
Ningbo lü Jing tongxianghui
　寧波旅京同鄉會
Ningbo yuye jingchaju 寧波
　漁業警察局
Ninghai 寧海
Ninghua 寧化
Nonggong yinhang 農工銀行

pengmin 棚民

qi 氣
qianzhuang 錢莊
Qingdao 青島
Qingsha 青沙
qingzhan 青鱣
qu 區
Qu Yingguang 屈映光
Quan Zhe gonghui 全浙
　公會
Qushan 衢山
quzhang 區長

Renhe Lodge 人和公所
Ruian 瑞安

Shanghai 上海
Shanghai shi shanghui 上海市
　商會

Shanghai yuye gaijin xuan-
　chuanhui 上海漁業改進
　宣傳會
Shanghai zong shanghui 上海
　總商會
shao laoban 少老闆
Shaoxing 紹興
shehuihua 社會化
Shen Changxin 沈昌鑫
shendong 紳董
Sheng county 嵊縣
Shengshan 嵊山
Shengsi 嵊泗
Shenjiamen 沈家門
Shenjiamen yuzhan gongsuo
　沈家門魚棧公所
Sheshan 佘山
Shi Meiheng 史美衡
Shi Renhang 史仁航
Shipu 石浦
Shishunxing 施順興
Shiye jihua 實業計劃
shuigui 水鬼
shuipiao 水票
Sijiao 泗礁
sishi jingli 司事經理
Song Ziwen 宋子文
Songjiang 松江
Songmen 松門
Suhua 蘇化
Sun Biaoqing 孫表卿
Sun Chuanfang 孫傳芳
Sun Guansheng 孫冠生
Sun Zhongshan 孫中山
Suyuan fa 訴願法

Taihe Lodge 泰和公所
Taizhou 台州

Taizhou Lodge 台州公所
Taizhou lü Chong tongxianghui 台州旅崇同鄉會
tanpai 攤派
Tanxu 灘滸
Tian Hou gong 天后宮
Tongfeng Lodge 同豐公所
Tongmenghui 同盟會
tongxingzheng 通行證
Tongzhao 桐照
tongzhi 統治
tu bang 土幫
tuowang 拖網

wai bang 外幫
Waihai shuijing ting 外海水警廳
Wang Jigao 王季高
Wang Jingwei 汪精衛
Wang Liansheng 王廉生
Wang Wentai 王文泰
Wang Wenxiang 王文襄
wangfan 網販
Wei Yuan 魏源
weichuan 喂船
Wenling 溫嶺
Wenling xian yuhui 溫嶺縣漁會
Wen-Tai shatian guanlichu 溫台沙田管理處
Wenzhou 溫州
Wenzhou lü Chong tongxianghui 溫州旅崇同鄉會
Wusong 吳淞
wuzei 烏賊

xia fan 下凡
xiangbaozhang 鄉保長
xianggui 鄉規

Xiangshan 象山
xiao duichuan 小對船
xiao huangyu 小黄魚
Xiaoyangshan 小洋山
Xiepu 蟹蒲
xieqian 謝錢
xie yang 謝洋
Xifeng 棲鳳
Xin Wenhuan 忻文焕
Xinxue huishe 新學會社
Xin yushang gongsuo 新魚商公所
xuncha 巡查
xunguan 巡官

yang fu 洋夫
Yangfugong 羊府宮
Yang Zhuan 羊偡
Yao Yongping 姚詠平
Ye Keliang 葉可樑
Yian Lodge 義安公所
Yihe Lodge 義和公所
Yikang qianzhuang 怡康錢莊
Yin County Fishing Association 鄞縣漁會
Yin dong waihai yuye hezuoshe 鄞東外海漁業合作社
Ying Mengqing 應夢卿
Yin county 鄞縣
Yin xian shuili ju 鄞縣水利局
Yongan Lodge 永安公所
Yongfeng Lodge 永豐公所
Yong River 甬江
Yongtai Lodge 永泰公所
Yu Qiaqing 虞洽卿
Yu Zhenxiang 虞禎祥
Yuansen yuhang 源森魚行
yu ba 漁霸
yu bang 漁幫

yu bing xia jiang 魚兵蝦將
Yu di 玉帝
yugai 漁改
yugang sheji weiyuanhui 漁港
　設計委員會
yu hang 魚行
yuhu 漁戶
yuhui 漁會
Yuhui fa 漁會法
Yuhui zanxing zhangcheng
　漁會暫行章程
Yuhui zanxing zhangcheng shi-
　xing xize 漁會暫行章程
　實行細則
yumin 漁民
Yumu si 漁牧司
yushang 魚商
Yushang gongsuo 魚商公所
Yutuan ju 漁團局
Yuxun jinyue 漁汛禁約
Yuyebu 漁業部
Yuye fa 漁業法
yuye gongsuo 漁業公所
yuye jianshe fei 漁業建設費
Yuye jingcha guicheng 漁業警
　察規程
Yuye ke 漁業課
Yuye shengchan zhihuibu 漁業
　生產指揮部
Yuye tiaoli 漁業條例
Yuye yintuan 漁業銀團
Yuye zhidaosuo 漁業指導所
yuye zibenjia 漁業資本家
yuye zuhe 漁業組合

za shui 雜稅
Zhang Boqi 張伯岐
Zhang Guifang 張桂芳
Zhang Jian 張謇

Zhang Liu 張鏐
Zhang Shenzhi 張申之
Zhang Xiaogeng 張曉耕
Zhang Yin 張寅
Zhang Yulu 張毓騄
Zhang Zhuzun 張柱尊
Zhang Zuolin 張作霖
Zhangshu 樟樹
zhangwang 張網
Zhao Cisheng 趙次勝
zhaowang 照網
Zhapu 乍浦
Zhejiang shengli jiazhong
　shuichan xuexiao 浙江省立
　甲種水產學校
Zhejiang shengli shuichanke
　zhiye xuexiao 浙江省立水產科
　職業學校
Zhejiang sheng yuye guanli
　weiyuanhui 浙江省漁業管理
　委員會
Zhejiang yutuan 浙江漁團
zhengdun 整頓
Zhenhai 鎮海
Zhe-Tai yuye gongsuo 浙台
　漁業公所
zhi 枳
zhishou 治首
Zhongguo shiye yinhang 中國
　實業銀行
Zhongguo tongshang yinhang
　中國通商銀行
Zhongguo yinhang 中國銀行
Zhonghua minguo shuichan
　xuehui 中華民國水產學會
Zhongjieshan 中街山
Zhou Jianyin 周監殷
Zhou Shitang 周世棠
Zhoushan qundao 舟山群島

Zhoushan zhuanshu 舟山專署   zizhi choubei weiyuan 自治籌備
zhu 柱   委員
Zhu Yunshui 朱云水   Zizhihui 自治會
zhushou 柱首   zong zhushou 總柱首
Zhuang Songfu 莊崧甫   Zuo Zongtang 左宗棠
zibenzhu 資本主

Notes

For complete author names, titles, and publication data for the works cited here in short forms, see the Works Cited, pp. 251–73.

Introduction

1. Liu Menglan, "Qugang yudeng," in *Daishan zhenzhi* 20.6b.
2. For discussions of China's contemporary environmental issues, see Economy, *The River Runs Black*; and Smil, *China's Past, China's Future*.
3. Jianguo Liu and Jared Diamond, "China's Environment in a Globalizing World," 6–7.
4. Worm et al., "The Impacts of Biodiversity Loss on Ocean Ecosystem Services."
5. Overall, China consumed massive amounts of fish, and total fishery production probably ranked among the highest in the world, as it does today. With China's large population, per capita fish consumption in early modern China was lower than in Japan and maritime countries in western Europe. However, in southeast China fish consumption was several times the countrywide average (Simoons, *Food in China*, 337–38).
6. Bolster, "Opportunities in Marine Environmental History."
7. Ostrom, *Governing the Commons*, 30–33; Ostrom et al., *Rules, Games, and Common-Pool Resources*, 6–8.
8. Ostrom et al., *Rules, Games, and Common-Pool Resources*, 10–12.
9. Ibid., 83.

10. Schlager, "Fishers' Institutional Reponses to Common-Pool Resource Dilemmas," 250–51.

11. Libecap, *Contracting for Property Rights*, 12–13, 73, 80; McEvoy, *The Fisherman's Problem*, 10–11; Ostrom, *Governing the Commons*, 3; Ostrom et al., *Rules, Games, and Common-Pool Resources*, 10; Schlager, "Fishers' Institutional Reponses to Common-Pool Resource Dilemmas," 250.

12. Feeny et al., "The Tragedy of the Commons"; McEvoy, "Toward an Interactive Theory of Nature and Culture," 226–28; Ostrom, *Governing the Commons*, 6–7; Ostrom et al., *Rules, Games, and Common-Pool Resources*, 15.

13. Ostrom, *Governing the Commons*, 51; Ostrom et al., *Rules, Games, and Common-Pool Resources*, 16–19.

14. Buoye, *Manslaughter, Markets, and Moral Economy*, chap. 3. For a gripping history of a particularly violent local society, see Rowe, *Crimson Rain*.

15. Perdue, *Exhausting the Earth*, 23, 167–70; Schoppa, *Song Full of Tears*, 160–63. In *Forest and Land Management in Imperial China*, Nicholas Menzies documents several instances of sustainable forestry management institutions under the Qing, such as imperial hunting preserves, temple forests, communal forests, and Cunninghamia planting. However, these fascinating examples were the exception rather than the norm.

16. Buoye, *Manslaughter, Markets, and Moral Economy*, 215.

17. In addition to settlement of legal cases, which is Huang's main concern, he notes that local officials relied on collaboration with local society for, among other functions, tax collection, public security, famine relief, and maintenance of public works; see Philip C. C. Huang, "Between Informal Mediation and Formal Adjudication," 288.

18. Scott, *Seeing Like a State*, 4.

19. Scoones, "Range Management Science and Policy," 51.

20. Smith, *Scaling Fisheries*.

21. Jennings et al., *Marine Fisheries Ecology*, 272–73; McNeill, *Something New Under the Sun*, 246–48.

22. Jennings et al., *Marine Fisheries Ecology*, 9–11. See also McEvoy, *The Fisherman's Problem*, 6.

23. Cushing, *The Provident Sea*, 112.

24. Iversen, *Living Marine Resources*, 247–48; Jackson et al., "Historical Overfishing and the Recent Collapse of Coastal Ecosystems," 635; Steneck and Carlton, "Human Alterations of Marine Communities," 446–48.

Chapter 1

1. Chen Ya-Qu and Xin-Qiang Shen, "Changes in the Biomass of the East China Sea Ecosystem"; Li Rongsheng, *Zhongguo shuichan dili*, 19–21; Qin Yunshan et al., *Geology of the East China Sea*, 1–15; Zhang Qilong and Wang Fan, "Zhoushan yuchang"; Zhejiang sheng haian he haitu ziyuan zonghe diaocha baogao bianxie weiyuanhui, *Zhejiang sheng haian he haitu ziyuan*, 452–53.

2. Zhejiang sheng shuichanzhi bianzuan weiyuanhui, *Zhejiang sheng shuichanzhi*, 59–62, 111–20.

3. Ya-Qu Chen and Shen Xin-Qiang, "Changes in the Biomass of the East China Sea Ecosystem," 96–100.

4. Zhejiang sheng shuichanzhi bianzuan weiyuanhui, *Zhejiang sheng shuichanzhi*, 110–15.

5. Huang Junming, "Daishan yuye lishi tedian tantao," 113; Zhao Yizhong, "Zhoushan yuye fazhan shi chutan," 105–7; Zhoushan yuzhi bianxie zu, *Zhoushan yuzhi*, 13–15.

6. Lavely et al., "Chinese Demography," 816.

7. Kuhn, *Soulstealers*, 41.

8. *Minguo Xiangshan xianzhi*, 4.2–3, *Zhenhai xianzhi* 41.41.

9. Shiba, "Ningpo and Its Hinterland," 401.

10. Ibid., 403. See also *Zhenhai xianzhi* 4.5.

11. Matsuura, *Chūgoku no kaizoku*, 112–13, 119, 128, 136, 141; Murray, *Pirates of the South China Coast*, 14–17.

12. Antony, *Like Froth Floating on the Sea*, 13–14, 17, 82, 97; Matsuura, *Chūgoku no kaizoku*, 8–9.

13. Ouyang Zongshu, *Haishang renjia*, 67, 121, 169.

14. Ibid., 136–37.

15. Ibid., 127–31.

16. Ibid., 16, 25.

17. Ibid., 142–44.

18. Ibid., 87–89, 169–70.

19. Ng, *Trade and Society*, 199.

20. Huang Junming, "Daishan yuye lishi tedian tantao," 114; Ouyang Zongshu, *Haishang renjia*, 27, 33; Zhao Yizhong, "Zhoushan yuye fazhan shi chutan," 108.

21. *Daishan zhenzhi* 5.1ab. See also Chen Musen et al., "Daishan xingshi tan," 167.

22. *Zhenhai xianzhi* 4.3

23. Ho, *Studies on the Population of China*, 156; Huang Junming, "Daishan yuye lishi tedian tantao," 116; Li Guoqi, *Zhongguo xiandaihua de quyu yanjiu*, 146–53; Zhao Yizhong, "Zhoushan yuye fazhan shi chutan," 109; Zhoushan yuzhi bianxie zu, *Zhoushan yuzhi*, 16–17.

24. Goodman, *Native Place, City, and Nation*, chap. 2.

25. Guo Zhenmin, *Shengsi yuye shihua*, 32–33; Jie Tianhai, "Caiyuan yuhang xingshuai ji," 29; Jin Tao, "'Shengsi yumin fengsu' kao," 90; Zhoushan yuzhi bianxie zu, *Zhoushan yuzhi*, 16.

26. Zhao Yizhong, "Zhoushan yuye fazhan shi chutan," 108–9.

27. China's macroregions, as Skinner explains, are not defined by administrative boundaries. Instead, they are socioeconomic units distinguished by physical geography and marketing patterns. Each macroregion is separated from others by mountains and integrated by river basins that carry goods between the regional core and the peripheries. See Skinner, "Regional Urbanization in Nineteenth-Century China," 211–20.

28. Zhang Qijun, *Zhejiang sheng shidi jiyao*, 79.

29. Osborne, "Highlands and Lowlands," 204; Shiba, "Environment Versus Water Control," 163.

30. Osborne, "Barren Mountains, Raging Rivers," 142, 153, 171–73, 182, 200, 202, 270.

31. For an insightful discussion of the relationship between the accumulation of silt in inland waters and the decline of freshwater fisheries in the Yangzi delta during the Qing, see Yi Lingling, *Ming Qing Changjiang zhong xia you yuye jingji yanjiu*, 243–44, 377–84. For sources detailing the environmental degradation evident in the Ningbo region's inland waters during the Qing, see Matsuda, "Min Shin jidai Sekkō Gin ken no suiri jigyō," 297–305.

32. Huang Junming, "Daishan yuye lishi tedian tantao," 118; Zhang Zhendong and Yang Jinsen, *Zhongguo haiyang yuye jianshi*, 112–13, 122; Zhoushan yuzhi bianxie zu, *Zhoushan yuzhi*, 159.

33. Zhoushan yuzhi bianxie zu, *Zhoushan yuzhi*, 109, 113.

34. Huang Junming, "Daishan yuye lishi tedian tantao," 118; Zhoushan yuzhi bianxie zu, *Zhoushan yuzhi*, 140, 143.

35. Shen Ligong, "Qiantan Gaoting de liuwang zuoye," 121.

36. Jiang Bin and Jin Tao, *Donghai daoyu wenhua yu minsu*, 85.

37. Ford, *An American Cruiser in the Far East*, 306.

38. On the history of the *dan* people during the Ming-Qing period, see Ouyang Zongshu, *Haishang renjia*, chap. 4.

39. These most prominent were the Taogongshan Xin, the Caojia Cao, the Shijiawan Shi, the Dayantou Dai, and the Yinjiawan Zheng lineages. For information on these lineages' economic pursuits, see the tables found in *Yin xian tongzhi*, 1: 368ab, 473a–74b, 511a, 517a. See also Lin Maochun and Wu Yuqi, *Yin xian yuye diaocha baogao*, 22.

40. Lin Maochun and Wu Yuqi, *Yin xian yuye diaocha baogao*, 25–26.

41. Jin Zhiquan, "Zhejiang yuye zhi xianzai ji jianglai zhi qushi," 47; Lin Maochun and Wu Yuqi, *Yin xian yuye diaocha baogao*, 22; Wu Zaisheng, "Shehui shenghuo," 129; Yao Yongping, "Daishan shuichan zhi diaocha," 8.

42. Geyao, "Zhoushan nanzi de chuantong zhiye," 814; Jin Zhiquan, "Zhejiang yuye zhi xianzai ji jianglai zhi qushi," 47.

43. Worcester, *Junks and Sampans of the Yangtze*, 1: 135. For an excellent anthropological investigation of gender relations in coastal areas of Fujian where many men migrated seasonally to Zhoushan's fishing grounds, see Friedman, *Intimate Politics*, 36–37.

44. Mann, "Women's Work in the Ningbo Area," 247. Fishing households from Zhejiang and Fujian spoke variants of the Yue and Hokkien dialects, but their gender division of labor resembled that of China's Hakka sub-ethnic group, in that it seems to have been "developed in response to male itinerancy and sojourning"; see Skinner, "Introduction."

45. Xiao Xiang, "Daishan zhuzhi ci"; recorded in *Daishan zhenzhi* 19.15a–b.

46. Worcester, *Junks and Sampans of the Yangtze*, 1: 131. See also Jin Zhiquan, "Zhejiang yuye zhi xianzai ji jianglai zhi qushi," 55; and Wu Zaisheng, "Shehui shenghuo," 128.

47. Tao Fusheng, "Jiazhi de yuye," 78.

48. Shenjiamen zhenzhi bianzuan lingdao xiaozu, *Shenjiamen zhenzhi*, 251.

49. Zhang Renyu, "Xinhai geming yilai de Dinghai shangye yanbian," 34–37.

50. Lin Maochun and Wu Yuqi, *Yin xian yuye diaocha baogao*, 5, 8–9; Wang Zongpei, "Zhongguo yanhai zhi yumin jingji," 113–14, 120–21.

51. Sources suggest that just 10–20 percent of fishing boats could come up with their own capital without reliance on loans; see Gan Yuli, "Jiang Zhe waihai yuye xiankuang," 215; and Jin Zhiquan, "Zhejiang yuye zhi xianzai ji jianglai zhi qushi," 57.

52. Mann, *Local Merchants and the Chinese Bureaucracy*, 174–78.

53. The use of the catch as collateral was analogous to the common practice called "selling the sprouts" (*maiqing*) in which farmers obtained capital by assigning moneylenders ownership of a portion of their sprouting crops. On this practice, see Mazumdar, *Sugar and Society in China*, 322–33.

54. *Daishan zhenzhi* 20.7b. See also Huang Junming and Zhang Mingquan, "Daishan yuhangzhan qianshuo," 43.

55. This aggregation of small-scale economic organizations resembles the type of cell-like or "plexus" market structure elaborated in the analysis of the Qing economy in Myers and Wang, "Economic Developments, 1644–1800," 586, 644. This conception of China's market economy also coincides with Mark Elvin's description in *The Retreat of the Elephants* (see esp. xxiii–xxiv, 113–14) of environmental exploitation characteristic of a "Chinese style" of premodern economic growth involving a "combination of small-unit initiative and all but unlimited facultative aggregation."

56. *Daishan zhenzhi* 5.2b.

57. Huang Junming and Zhang Mingquan, "Daishan yuhangzhan qianshuo," 43.

58. Chen Guoqiang and Cai Yongzhe, *Chongwu renleixue diaocha*, 8.

59. William Cronon describes how similar reliance on credit allowed lumber businesses to survive seasonal downturns in America's Great Lakes region during the nineteenth century; see Cronon, *Nature's Metropolis*, 168–69.

60. Elvin, *Retreat of the Elephants*, 82–83.

61. Ibid., xviii.

62. Jin Zhiquan, "Zhejiang yuye zhi xianzai ji jianglai zhi qushi," 51.

63. Chen Guoqiang and Cai Yongzhe, *Chongwu renleixue diaocha*, 63–64; Jie Tianhai, "Caiyuan yuhang xingshuai ji," 29; "Nan Shi no suisan," 39; Shenjiamen zhenzhi bianzuan lingdao xiaozu, *Shenjiamen zhenzhi*, 252.

64. Guo Zhenmin, *Shengsi yuye shihua*, 171. Fish-processing merchants in Daishan sometimes did business importing these supplies as well; see Huang Junming and Zhang Mingquan, "Daishan yuhangzhan qianshuo," 37–38; and Jin Li, "Hengjie yushi shihua," 173.

65. "Nan Shi no suisan," 190; *Daishan zhenzhi* 5.2ab, 20.7ab; He Yunyu and Wei Xiyan, "Jianguo qian Chongwu yuye gaikuang," 8–9; Huang Junming and Zhang Mingquan, "Daishan yuhangzhan qianshuo," 43; *Kōsō shō Sekkō shō suisangyō chōsa hōkoku*, 26; Tōa dōbunkai, *Shina shōbetsu zenshi*, 14: 548; Wu Zaisheng, "Shehui shenghuo," 129.

66. For similar pricing policies in a fishing village in the vicinity of Hong Kong during the 1950s, see King, "Pricing Policy in a Chinese Fishing Village."

67. This term is drawn from Howell, *Capitalism from Within*, 48–49.

68. Acheson, "Anthropology of Fishing," 282. Barbara E. Ward ("Chinese Fishermen in Hong Kong," 276, 282) notes similar credit relations in Hong Kong, where boat-dwelling *dan* people and the dealers who provided them with credit often came from different ethnic groups.

69. Brokers performed these functions for small-scale producers throughout the Chinese economy; see Mann, *Local Merchants and the Chinese Bureaucracy*, 176–77.

70. Ibid., 175.

71. Guo Zhenmin, *Shengsi yuye shihua*, 178–80; Zhoushan shizheng wenshi he xuexi weiyuanhui, *Zhoushan haiyang yu wenhua*, 67.

72. During the late Qing, yellow croaker runs attracted dried biscuit peddlers from Huangyan county in Taizhou to Qushan Island, where they rented buildings to sell their wares before returning home at the end of fishing season. In the early twentieth century, some of these merchants settled down to open up shops permanently on the island; see Gu Zongjian, "Weijingtan yinggao," 169–70.

73. Guo Zhenmin, *Shengsi yuye shihua*, 178–80; Zhoushan shizheng wenshi he xuexi weiyuanhui, *Zhoushan haiyang yu wenhua*, 67.

74. Fu Guozhang, "Yuhang yu bingxian," 25.

75. *Daishan zhenzhi* 5.2b–3a, 20.8a; Jin Li, "Hengjie yushi shihua," 173.

76. For discussions of icehouses, see *Yin xian tongzhi*, 1: 44; Chen Yuxin, "Zhoushan bingchang," 928–29; and Worcester, *Junks and Sampans of the Yangtze*, 1: 127–28.

77. Fortune, *Three Years' Wanderings in the Northern Provinces of China*, 94.

78. Fu Guozhang, "Yuhang yu bingxian," 25–28; Himeda, "Chūgoku kindai gyogyō shi no hitokoma," 89–92; Huang Junming and Zhang Mingquan, "Daishan yuhangzhan qianshuo," 45; Zhoushan shizheng wenshi he xuexi weiyuanhui, *Zhoushan haiyang yu wenhua*, 58–60; Zhoushan yuzhi bianxie zu, *Zhoushan yuzhi*, 246–48, 254. This credit system sparked riots in Ningbo during 1858 when a currency shortage caused native banks to rediscount the certificates so steeply that fishers received only half of their face value. For an analysis of this incident, see Himeda, "Chūgoku kindai gyogyō shi no hitokoma," 86–88.

79. Jin Zhiquan, "Zhejiang yuye zhi xianzai ji jianglai zhi qushi," 58.

80. Lin Maochun and Wu Yuqi, "Yin xian yuye zhi diaocha," 17. See also "Zhejiang qu yuye guanli ju cheng," July 1931, IMH 17-27 135-3.

81. For extensive research tracing the use of ice boats to the late Ming dynasty, see Qiu Zhonglin, "Bingxian chuan yu xianyu hang."

82. Worcester, *Junks and Sampans of the Yangtze*, 1: 126.

83. Zhao Yizhong, "Zhoushan de bingxian shang he yuhang jianxi," 66.

84. Fu Guozhang, "Yuhang yu bingxian," 26.

85. Zhao Yizhong, "Zhoushan yuye fazhan shi chutan," 66.

86. Shanghai bowuguan, Tushu ziliao shi, *Shanghai beike ziliao xuanji*, 417.

87. Jin Zhiquan, "Zhejiang yuye zhi xianzai ji jianglai zhi qushi," 45.

88. "Huyu banshichu cheng," July 13, 1931, IMH 17-27 135-3.

89. During the nineteenth and early twentieth centuries, Chinese state authorities allowed fishers and fish merchants to purchase salt for preserving and processing the catch directly from salt-producing households at discounted tax rates. Fishery producers often sold leftover salt illicitly, giving rise to accusations of smuggling. See *Daishan zhenzhi* 4.9b–11b; *Dinghai xianzhi* 4.7a–8b; and Li Shihao and Qu Ruoqian, *Zhongguo yuye shi*, 212–14.

90. Fu Guozhang, "Yuhang yu bingxian," 27; Guo Zhenmin, *Shengsi yuye shihua*, 169.

91. "Wang Jihong cheng," June 26, 1931, IMH 17-27 135-3. See also Fu Guozhang, "Yuhang yu bingxian," 26–27; and Zhoushan yuzhi bianxie zu, *Zhoushan yuzhi*, 249–54.

92. "Nan Shi no suisan," 36.

93. "Wang Jihong cheng," June 26, 1931, IMH 17-27 135-3. Charging a premium when converting sycee taels into silver dollars was a common practice among native banks; see Linsun Cheng, *Banking in Modern China*, 141.

94. Jin Zhiquan, "Zhejiang yuye zhi xianzai ji jianglai zhi qushi," 45.

95. Robert Marks (*Tigers, Rice, Silk, and Silt*, 174) and Sucheta Mazumdar (*Sugar and Society in China*) find analogous marketing relations in the sugar industry of south China during the Qing, and Elvin (*Retreat of the Elephants*, 214–15) points to parallel marketing systems in the Lower Yangzi cotton industry. In each instance, capital that originated from regional economic centers reached small-scale household producers in the form of credit provided by merchant middlemen.

Chapter 2

1. Kuhn, "Toward a Historical Ecology of Chinese Migration."
2. Zhao Yizhong, "Zhoushan de yuye gongsuo," 66.
3. McEvoy, *The Fisherman's Problem*, 96. See also McGoodwin, *Crisis in the World's Fisheries*, 123.
4. *DHZB*, May 21, 1934, 2–3; June 6, 1934, 1–2. See also Jin Li, "Hengjie yushi shihua," 171–72.
5. Xu Shihe, "Jiang Zhe liang sheng zhi yuye," 1.
6. *Kōsō shō Sekkō shō suisangyō chōsa hōkoku*, 17.
7. Li Yaohui, "Jiangsu zhi yuye gaikuang," 110.
8. Worcester, *Junks and Sampans of the Yangtze*, 1: 135.
9. Acheson, "Anthropology of Fishing," 281.
10. Weller, *Discovering Nature*, 39–41, 84–85, 106.
11. Fang Changsheng, ed., *Zhejiang sheng minjian wenxue jicheng*, 19.
12. Xu Bo, *Zhoushan fangyan yu Donghai wenhua*, 241–43.
13. Fang Changsheng, ed., *Zhejiang sheng minjian wenxue jicheng*, 36.
14. Jiang Bin and Jin Tao, *Donghai daoyu wenhua yu minsu*, 439, 443–44.
15. *Xin xiu Yin xianzhi* 13.3a–4a. Other secondary sources state that the temple honored a Jin dynasty (266–316) magistrate from Yin county; see, e.g., Wang Rongguo, *Haiyang shenling*, 271.
16. Wang Rongguo, *Haiyang shenling*, 271.
17. See Watson, "Standardizing the Gods."
18. Gong Yu and Huang Zhiguo, "Zhoushan de 'Mazu miao,'" 787.
19. Reprinted in Yu Fuhai, ed., *Ningbo shizhi waibian*, 1002.
20. Dongshajiao's Yang Fu Temple was first constructed in 1761 and later renovated during the Guangxu period (1875–1908); see *Daishan zhenzhi* 10.8a.
21. Jiang Bin and Jin Tao, *Donghai daoyu wenhua yu minsu*, 232–35.
22. For a discussion of technological externalities, see Schlager, "Fishers' Institutional Reponses to Common-Pool Resource Dilemmas," 252–53.
23. Fang Changsheng and Wang Daoxing, *Zhoushan fengsu*, 85–86; Fang Changsheng, *Zhoushan minsu wenxue yanjiu*, 61.
24. Schlager, "Fishers' Institutional Responses to Common-Pool Resource Dilemmas," 251.
25. Duara, *Culture, Power, and the State*, 30–35.

26. See Goodman, *Native Place, City, and Nation*, 41–46; Golas, "Early Ch'ing Guilds"; Rowe, *Hankow*; and Naquin, *Peking*, 598–99.

27. *Dinghai xianzhi* 3.3b; Zhao Yizhong, "Zhoushan de yuye gong-suo," 67.

28. According to the table of fishing lodges in the Dinghai county gazetteer, 8 fishing lodges were established before 1850, 35 between 1850 and 1911, and 38 from 1911 to 1925; see *Dinghai xianzhi* 3.3a–5b.

29. Fishermen and merchants who migrated from the Ningbo region to the fishing port of Shenjiamen gathered to worship at Yang Fu Temple, which also housed the offices of the Yongan Lodge. The Fujian group's Bamin Lodge in Shenjiamen occupied the Empress of Heaven Temple (Tian Hou gong) dedicated to Mazu. The Taizhou Fishing Lodge also used a temple in Daishan as its headquarters before purchasing property for its own building in 1925. See *Daishan zhenzhi* 9.5a; *Dinghai xianzhi* 2.13b; Shenjiamen zhenzhi bianzuan lingdao xiaozu, *Shenjiamen zhenzhi*, 258.

30. Li Shiting, "'Sucheng' xiaokao," 809.

31. Zhang Jian, "Yumin 'xie yang,'" 125–27. See also Zhoushan shizheng wenshi he xuexi weiyuanhui, *Zhoushan haiyang long wenhua*, 85.

32. De Groot, *The Religious System of China*, 5: 525.

33. "Yuchuan zhangcheng," in Zhenhai xianzhi bianzuan weiyuanhui, *Zhenhai xianzhi*, 991.

34. Naquin, *Peking*, 616. See also Goodman, *Native Place, City, and Nation*, 91–92, 96.

35. *Daishan zhenzhi* 5.2b.

36. Ibid., 9.3ab–4b. For a discussion of the development of fish-processing enterprises in Daishan, see Ding Fanglong and Guan Baoren, "Jiefang qian Daishan shangye gaikuang," 26. If a fishing lodge's membership grew too large, it could divide into multiple organizations. Sojourning fishermen from Tongzhao village in Fenghua county established the Yihe Lodge on Daishan Island in 1798. By the early nineteenth century, the number of boats from Tongzhao that came to fish in Daishan had greatly increased. For this reason, the Yihe Lodge split to form a second organization called the Yian Lodge in 1813. After this division, the two lodges rented a building to serve as their shared headquarters. The Yihe Lodge and the Yian Lodge did not establish separate headquarters until 1918, when each purchased property in Dongshajiao to construct its own office; see *Daishan zhenzhi* 9.4b. The Old Fish Merchants Lodge and New Fish Merchants Lodge shared a

rented building until 1908, when the latter finally broke off and purchased its own headquarters (ibid., 9.3b–4a).

37. Jin Zhiquan, "Zhejiang yuye zhi xianzai ji jianglai zhi qushi," 55; Li Shihao and Qu Ruoqian, *Zhongguo yuye shi*, 96; Zhao Yizhong, "Zhoushan de yuye gongsuo," 73–74.

38. Shen Guangshi, "Jiang Zhe yuye shicha baogao," 170.

39. *Daishan zhenzhi* 20.6b. See also Zhao Yizhong, "Zhoushan de yuye gongsuo," 67.

40. "Yin xian Dongxiang Yongan gongsuo leshi bei," in Zhejiang sheng shuichanzhi bianzuan weiyuanhui, *Zhejiang sheng shuichanzhi*, 1147.

41. Zhao Yizhong, "Zhoushan de yuye gongsuo," 73–74.

42. *Daishan zhenzhi* 20.6b; Takumukyoku, *Chūnan Shina hōmen ni okeru suisan jijō*, 190. At least one source indicates that lodge organizations set opening prices, but auction bidding determined the final price. See *DHZB*, June 6, 1934, 1; and Zhao Yizhong, "Zhoushan de yuye gongsuo," 66.

43. Chen Qingcui, "Ji Juexisuo zhi yuxun" (ca. 1930s), in Chen Shanqing, ed., *Dongchen cunzhi waibian*, 31.

44. *Dinghai tingzhi* 20.40a.

45. Kuhn, *Rebellion and Its Enemies in Late Imperial China*, 213.

46. Rankin, *Elite Activism and Political Transformation in China*, 152–53.

47. *Dinghai xianzhi* 4.7a; Li Shihao and Qu Ruoqian, *Zhongguo yuye shi*, 33–37.

48. Zhu Yunshui, "Zhejiang yutuan yan'ge shi" 75–76. On the impact of the Sino-Japanese War in Zhejiang, see Rankin, *Elite Activism and Political Transformation in China*, pp. 165–66.

49. Li Shihao and Qu Ruoqian, *Zhongguo yuye shi*, 35.

50. Ibid., 38–39; Zhu Yunshui, "Zhejiang yutuan yan'ge shi," 101.

51. *Daishan zhenzhi* 5.7ab.

52. Imperial Maritime Customs, *Reports on the Trade of the Treaty Ports of China: Taichow*, 23.

53. Imperial Maritime Customs, *Decennial Reports on the Trade, Navigation, Industries, etc.*, 62.

54. Kobayashi, *Shina no janku*, 42; Rankin, *Elite Activism and Political Transformation in China*, 43, 59; Zhu Zhengyuan, *Zhejiang sheng yanhai tushuo*, 28b. Often these pirates forced fishing boats to pay protection money for "pirate permits"; see Li Shihao, *Zhongguo haiyang yuye xianzhuang ji qi jianshe*, 203–4.

55. *Daishan zhenzhi* 5.7b–8a

56. Ibid., 5.3b–4a, 20.6ab; Zhu Zhengyuan, *Zhejiang sheng yanhai tu-shuo*, 40a.

57. *Daishan zhenzhi* 20.6b–7a.

58. Zhou Qingsen, "Yangsheng shi," in *Daishan zhenzhi* 19.17b–18a.

59. The year 1906 does not appear to have experienced low levels of precipitation or Yangzi River runoff. On the other hand, it is possible that the poor fishing season resulted from weak summer monsoon winds that year. These environmental conditions hamper the distribution of nutrient-rich runoff from the Yangzi River in China's coastal waters, which has a negative effect on fishery production; see Qiu Yongsong et al., "Runoff- and Monsoon-Driven Variability of Fish Production in East China Seas," 31–32. For indices of summer monsoon variability, see Ge Quansheng et al., "1736 nian yilai Changjiang zhong xia you meiyu bianhua," 2797; and Guo Qiyun et al., "1873–2000 nian Dongya xiajifeng bianhua de yanjiu."

60. See Kuhn, *Rebellion and Its Enemies in Late Imperial China.*

61. According to one source, the first fishermen to come to Huanglong were migrants from Wenzhou and Taizhou, who fished there only on a seasonal basis. After fishermen from the Ningbo region heard about Huanglong's fishing grounds in the 1850s, they flocked to the island and soon crowded out the original sojourners from southern Zhejiang (Jiang Bin and Jin Tao, *Donghai daoyu wenhua yu minsu*, 39).

62. Jin Gou, "Huanglong dao lishi yan'ge," 90–91; Zhu Jin'gou, "Qing-mo he minguo shi Huanglong dao shang shou hang chuan," 83. The director of the Huanglong group's Beiyang Lodge, Liu Tongjiao, was reportedly an examination degree–holder and the owner of a silk fabric wholesale shop in the port of Dongmen in Xiangshan county (Zhu Jin'gou, "Qingmo he minguo shi Huanglong dao shang shou hang chuan," 22–23).

63. Jin Gou, "Huanglong dao lishi yan'ge," 31.

64. Zhu Jin'gou, "Qingmo he minguo shi Huanglong dao shang shou hang chuan," 50–51.

65. Jin Gou, "Huanglong dao lishi yan'ge," 23–24.

66. Ibid., 83.

67. Ying Mengqing, "Fenghua yumin gansidui canjia guangfu Hangzhou," 71. Ying's narrative was first published in 1981 as part of a collection of oral histories about the 1911 Revolution in Zhejiang. They were later reproduced in edited form as part of another collection. I refer to the earlier edition only for portions omitted from the later version.

68. Ying joined the Revolutionary Alliance while studying in Japan. After returning to China in 1907, Ying served as an instructor at various educational institutions in Shanghai and eventually became a clerk at the Revolutionary Alliance's central headquarters in the city during the fall of 1911 (Fenghua shizhi bianzuan weiyuanhui, *Fenghua shizhi*, 881–82).

69. Ying Mengqing, "Fenghua yumin gansidui canjia guangfu Hang-zhou," 71–72.

70. Ying Mengqing, "Fenghua yumin canjia guangfu Hangzhou gansi-dui ji," 187.

71. Ying Mengqing, "Fenghua yumin gansidui canjia guangfu Hang-zhou," 72.

72. Sources indicate that even during the early 1940s all directors of fishing organizations in Xifeng came from the Shen lineage; see "Zhe-jiang sheng Fenghua xian yuye chanxiao hezuoshe zhiyuan biao," August 1946; and "Xifeng yuye shengchan hezuoshe zhiyuan biao," October 1946: both in SMA Q462-117.

73. Ying Mengqing, "Fenghua yumin gansidui canjia guangfu Hang-zhou," 72.

74. Ibid.

75. Ibid., 73.

76. Ibid.

77. Ibid., 74.

78. Ibid. The New Learning Company was a locus of interaction for progressive elites from Fenghua and students connected with the Revolutionary Alliance who had recently returned from studying in Japan in the years leading up to the 1911 Revolution. The Fenghua elites Zhou Shitang and Sun Biaoqing established the bookstore in the early twentieth century to publish translations of books dedicated to foreign learning. Most of the elites associated with the publishing company had studied at the Longjin Academy, which was the center for "new learning" in Fenghua. Many progressive students had studied at the Longjin Academy. The New Learning Company published Chinese translations of Japanese books, making it a gathering place for returned students who joined the Revolutionary Alliance. See Sun Biaoqing and Mao Yihu, "Xinxue huishe ji qita," 55–57.

79. Fenghua shizhi bianzuan weiyuanhui, *Fenghua shizhi*, 877; *Daishan zhenzhi* 5.10b.

80. Ying Mengqing, "Fenghua yumin canjia guangfu Hangzhou gan-sidui ji," 190.

81. Ying Mengqing, "Fenghua yumin gansidui canjia guangfu Hang-zhou," 74–75.

82. Ibid., 75.

83. This point is stressed in Libecap, *Contracting for Property Rights*, 116.

84. Susan Mann (*Local Merchants and the Chinese Bureaucracy*, 12–13) uses this term, derived from the work of Max Weber, to connote the public services, such as tax collection, that merchant communities performed on behalf of the state.

85. Perdue, "Lakes of Empire," 123–25.

86. For a comparative analysis of this issue, see Schlager, "Fishers' Institutional Reponses to Common-Pool Resource Dilemmas."

Chapter 3

1. Schneider, *Biology and Revolution in Twentieth-Century China*, 3.

2. This vision of environmental management informed the modern Chinese state's water management and forestry policies as well; see Pietz, *Engineering the State*, 28, 36, 119, 122; and Songster, "Cultivating the Nation in Fujian's Forests," 454, 462, 469.

3. Dinghai xianzhi bianzuan weiyuanhui, *Dinghai xianzhi*, 108. Population statistics for the years prior to 1920 can be found in *Dinghai xianzhi* 1.18b–21a. As with all Chinese population statistics, these figures are not exact, but they convey the overall trend toward population growth in the Zhoushan Archipelago during the early twentieth century.

4. Chen Musen et al., "Daishan xingshi tan," 170–71; Zhao Yizhong, "Zhoushan yuye fazhan shi chutan," 108; Zhoushan yuzhi bianxie zu, *Zhoushan yuzhi*, 16.

5. *Daishan zhenzhi* 5.1a.

6. Ibid., 5.1b, 3b; Kobayashi, *Shina no janku*, 217; Takumukyoku, *Chūnan Shina hōmen ni okeru suisan jijō*, 190. The main occupation of Daishan Island's permanent residents was the production of sea salt used for fish processing and exported for sale in other regions. Agriculture did not make up a major part of local incomes, since Daishan had little arable land. The island depended on imports for most of its grain reserves. See *Daishan zhenzhi* 5.1ab, 16.34b; and Kobayashi, *Shina no janku*, 215.

7. *Dinghai xianzhi* 16.34a.

8. Ibid., 18.2; Ding Fanglong and Guan Baoren, "'Penglai shi jing' tan," 25; Huang Junming, "Daishan yuye lishi tedian tantao," 120; Huang

Junming and Zhang Mingquan, "Daishan yuhangzhan qianshuo," 36–37; Jin Li, "Hengjie yushi shihua," 172–73.

9. See Rawski, *Economic Growth in Prewar China*, chap. 1.

10. Xu Bin, "Penglai hangdu shu Nanpu," 154–55.

11. *Dinghai xianzhi* 5.34a; Zhang Renyu, "Xinhai geming yilai de Dinghai shangye yanbian," 36; Mann, "Women's Work in the Ningbo Area," 250.

12. Rawski, *Economic Growth in Prewar China*.

13. "Nan Shi no suisan," 36, 39; Tōa dōbunkai, *Shina shōbetsu zenshi*, 14: 556.

14. Zhejiang sheng shuichanzhi bianzuan weiyuanhui, *Zhejiang sheng shuichanzhi*, 897. Shenjiamen had only one native bank in 1920, but eighteen were doing business there by 1932 (Shenjiamen zhenzhi bianzuan lingdao xiaozu, *Shenjiamen zhenzhi*, 362). The first native banks on Daishan Island opened in 1927. Four more opened by the end of the decade; see Tu Hengting, "Minguo shiqi Daishan jinrongye gaikuang," 14.

15. Zhejiang sheng shuichanzhi bianzuan weiyuanhui, *Zhejiang sheng shuichanzhi*, 60–61, 120.

16. Takumukyoku, *Chūnan Shina hōmen ni okeru suisan jijō*, 260.

17. Tōa dōbunkai, *Shina shōbetsu zenshi*, 14: 36; Zhao Yizhong, "Zhoushan de bingxian shang he yuhang jianxi," 72. For a general discussion of the growth of native banks in the Ningbo region during the Qing dynasty, see Mann, "Finance in Ning-po."

18. Guo Jizhong, "Kanmen san qianzhuang," 45–47.

19. Tu Hengting, "Minguo shiqi Daishan jinrongye gaikuang," 14–15.

20. "Zhongguo tongshang yinhang guanyu chazhang baogao ji qiankuan cunshou Daishan duihuanchu bing renshi waidiao cizhi jiaxin baozheng deng yu Dinghai banshichu wanglai wenshu," March 17, 1934, SMA 281-1-545. See also Shenjiamen zhenzhi bianzuan lingdao xiaozu, *Shenjiamen zhenzhi*, 362.

21. Sheng Guanxi, "Jindai Zhoushan de diandangye," 19. The first pawnshop on Daishan had opened during the Guangxu period (1875–1908), but six more started business between 1918 and 1929; see Tu Hengting, "Minguo shiqi Daishan jinrongye gaikuang," 13.

22. Sheng Guanxi, "Jindai Zhoushan de diandangye," 20.

23. On regional inequalities resulting from concentration of capital in highly developed regions that extract resources and labor from marginal maritime communities, see Prattis, "Modernization and Modes of Production in the North Atlantic."

24. "Nan Shi no suisan," 39.

25. Tu Hengting, "Minguo shiqi Daishan jinrongye gaikuang," 14–16.

26. Hoffmann, "Economic Development and Aquatic Ecosystems in Medieval Europe," 652.

27. "The Fishing Industry in Kiangsu," 836.

28. "Yin xian Dongxiang Yongan gongsuo leshi bei," in Zhejiang sheng shuichanzhi bianzuan weiyuanhui, *Zhejiang sheng shuichanzhi*, 1148. Throughout the Republican period, pirates hid out on the island of Changtushan directly to the east of Daishan and the islands of Sijiao and Huaniaoshan to the north, preying on passing boats and shipping; see Kobayashi, *Shina no janku*, 40.

29. Mazumdar, *Sugar and Society in China*, 319, 322.

30. Antony, *Like Froth Floating on the Sea*, 13–14, 82, 97; Matsuura, *Chūgoku no kaizoku*, 8–9.

31. *DHZB*, October 14, 1934, 3; October 16, 1934, 2.

32. *Daishan zhenzhi* 5.8ab.

33. Zhuang Jingzhong (Songfu), *Qiuwo shanren nianpu*, 4b, 7a. Zhuang Songfu (1860–1940) passed the county examinations in 1890 and became director of the Longjin Academy alongside Jiang Beiming in 1903. Jiang also helped to smooth over popular indignation that arose when Zhuang allowed the destruction of several wooden idols to construct another school at a temple in Fenghua. Zhuang took over management of the New Learning Company in 1905 and joined the Revolutionary Alliance in 1908. Following the 1911 Revolution, Zhuang served in the Zhejiang provincial government as head of the Department of Finance, director of the Salt Gabelle, and a member of the provincial assembly. In addition to Jiang Beiming, Zhuang also had close connections with Ying Mengqing, with whom he founded a tree-farming enterprise in Zhejiang's Lin'an county in 1920. Zhuang Songfu also cooperated with the Yongfeng Lodge's director Zhang Shenzhi in formulating water conservancy plans in Fenghua and Yin county. See Mao Yihu, "Zhuang Songfu de yisheng"; Wang Weimin, "Huiyi Zhuang Songfu xiansheng"; Zhuang Jingzhong (Songfu), *Qiuwo shanren nianpu*, 8b–9a; and Fenghua shizhi bianzuan weiyuanhui, *Fenghua shizhi*, 876–877.

34. *Daishan zhenzhi* 5.9a–10a.

35. The agreement also prohibited fishermen from bringing weapons to Daishan. Lodge leaders were to turn fishermen who were discovered secretly hoarding weapons over to the local authorities. Any fisherman who possessed licensed arms had to give them to his native-place lodge

for safekeeping until he returned home at the end of fishing season (ibid., 5.10b–12a).

36. Inspectors charged with favoritism or covering up transgressions by fishermen from their native-place would be removed from office. Local government authorities also threatened to take action against lodge leaders if they failed to discipline their compatriots (ibid., 5.11b).

37. Walker, "Meiji Modernization, Scientific Agriculture."

38. Boorman, *Biographical Dictionary of Republican China*, 1: 35; Guo Zhenmin, *Shengsi yuye shihua*, 42; Shanghai yuyezhi bianzuan weiyuanhui, *Shanghai yuyezhi*, 545. For discussions of Zhang's modernization programs in his own native place of Nantong, Jiangsu, see Köll, *From Cotton Mill to Business Empire*; and Shao, *Culturing Modernity*.

39. Zhang Jian, "Shangbu toudeng guwenguan Zhang zicheng benbu chouyi yanhai ge sheng yuye banfa wen," *DFZZ*, March 1906, 20.

40. "Shangbu toudeng guwen Zhang dianzhuan Jian zicheng Liang-Jiang zongdu Wei yichuang Nanyang yuye gongsi wen," *DFZZ*, September 1904, 147.

41. Ibid., 146–49.

42. "Jiang-Zhe yuye gongsi jianming zhangcheng," *DFZZ*, December 1904, 189.

43. "Shangbu toudeng guwen Zhang dianzhuan Jian zicheng Liang-Jiang zongdu Wei yichuang Nanyang yuye gongsi wen," *DFZZ*, September 1904, 148.

44. Ibid.

45. "Shangbu toudeng guwenguan Zhang zicheng benbu chouyi yanhai ge sheng yuye banfa wen," *DFZZ*, March 1906, 24.

46. "Jiang-Zhe yuye gongsi jianming zhangcheng," *DFZZ*, December 1904, 189.

47. "Shangbu toudeng guwen Zhang dianzhuan Jian zicheng Liang-Jiang zongdu Wei yichuang Nanyang yuye gongsi wen," *DFZZ*, September 1904, 149.

48. For an annual fee of one or two *yuan* depending on their size, boats that joined the Fishing Association received a registration flag from the Jiangsu-Zhejiang Fishing Company; "Jiang-Zhe yuye gongsi yuhui zhangcheng," *DFZZ*, March 1904; "Shangbu toudeng guwenguan Zhang zicheng benbu chouyi yanhai ge sheng yuye banfa wen," *DFZZ*, March 1906, 22. See also Zhoushan shi dang'anguan, *"Shenbao" Zhoushan shiliao huibian*, 51–52. On the history of the *lijin* tax and ways merchants sought to avoid the payment, see Mann, *Local Merchants and the Chinese Bureau-*

cracy, chaps. 6–7. For the role of Shanghai native-place organizations in collecting of *lijin*, see Goodman, "The Native Place and the City," 147–57.

49. *SB*, July 6, 1926, 15. See also Li Shihao and Qu Ruoqian, *Zhongguo yuye shi*, 42, 154–56.

50. Huang Zhenshi, "Jiu Shanghai de yushi," 221.

51. "Shangbu toudeng guwen Zhang dianzhuan Jian zicheng Liang-Jiang zongdu Wei yichuang Nanyang yuye gongsi wen," *DFZZ*, September 1904, 112.

52. Gu Mingsheng, trans., *Shuichanxue xinbian*, 1–2.

53. "Shangbu toudeng guwenguan Zhang zicheng benbu chouyi yanhai ge sheng yuye banfa wen," *DFZZ*, March 1906, 26–27.

54. Zhang Liu (1882–1925) was born in Jiangsu's Jiading county but moved with his father to study at academies in Jinhua, Yiwu, and other locations in Zhejiang before leaving to study in Japan (Shanghai yuyezhi bianzuan weiyuanhui, *Shanghai yuyezhi*, 546–47). In 1916, the Zhejiang provincial government opened another fishery school in Linhai county, which later moved to Dinghai in 1927 (Li Shihao and Qu Ruoqian, *Zhongguo yuye shi*, 131, 134–35).

55. Wang Tang, "Duiyu zhengdun Zhongguo yuye zhi guanjian," *SB*, June 22, 1924 (supplement), 2.

56. Li Shihao, *Zhongguo haiyang yuye xianzhuang ji qi jianshe*, 4–6; Li Shihao and Qu Ruoqian, *Zhongguo yuye shi*, 170–77.

57. Liu Tongshan and Xu Jibo, "Zhongguo yanhai yuye yu yumin shenghuo," 93, 95; Huang Zhenshi, "Jiu Shanghai de yushi," 229.

58. Zhang Liquan, "Shanghai haichanwu shichang zhi qushi," 28.

59. Wang Tang, "Duiyu zhengdun Zhongguo yuye zhi guanjian," *SB*, June 5, 1924 (supplement): 3.

60. See Ninohei, *Nihon gyogyō kindai shi*, 105–11.

61. After his return to China, Li had advised the central government's Ministry of Agriculture and Forestry on fishery affairs and held an administrative post at the Jiangsu Provincial Fishery School (Shanghai yuyezhi bianzuan weiyuanhui, *Shanghai yuyezhi*, 547; Zhejiang sheng shuichanzhi bianzuan weiyuanhui, *Zhejiang sheng shuichanzhi*, 1008–9). On the Dinghai Fishery Training Institute, see Zhoushan yuzhi bianxie zu, *Zhoushan yuzhi*, 335–336.

62. The central government's Ministry of Agriculture and Commerce first directed the Zhejiang provincial government to make preparations

for the creation of fishery training institutes in 1914 (Zhoushan shi dang'anguan, *"Shenbao" Zhoushan shiliao huibian*, 149).

63. The Training Institute had a hard time convincing fishermen to come to its headquarters to engage in study. Many of the new types of fishing gear that the institute encouraged, such as nets made from cotton fibers, were not as effective as fishermen's traditional techniques (ibid., 159, 161–62, 167). According to the gazetteer of Dinghai country, the number of fishermen that could attend the Dinghai Fishery Training Institute's lectures was limited to a mere twenty; see *Dinghai xianzhi* 5.7a–8a; and also Li Shihao and Qu Ruoqian, *Zhongguo yuye shi*, 25, 101.

64. Li Shihao and Qu Ruoqian, *Zhongguo yuye shi*, 25. For biographical information on Wang Wentai, see "Huiyuan lu," 112, in *Zhonghua minguo shuichan xuehui huibao* (1934): SMA Y4-1-225.

65. "Wang Wentai chen song Haizhou yuye jihuashu cheng gao," February 20, 1920, in Zhongguo dier lishi dang'anguan, *Zhonghua minguo shi dang'an ziliao huibian*, vol. 1, 3: 682.

66. Ibid., 682–83.

67. Guan Pengwan, *Shuichanxue da yi*, 49. See also Gu Mingsheng, trans., *Shuichanxue xinbian*, 1, 12.

68. Gu Mingsheng, trans., *Shuichanxue xinbian*, 1. This view was not universal. Because of the abundance of China's saltwater fishery resources and the underdevelopment of its fishing fleet, many fishery specialists maintained that marine aquaculture was not of great importance; see Zhou Jianyin and Yu Huaxian, *Zhongdeng shuichanxue*, 82.

69. Guan Pengwan, *Shuichanxue da yi*, 50.

70. Gu Mingsheng, trans., *Shuichanxue xinbian*, 22–23; Zhang Zhuzun, "Fazhan Zhejiang shuichan jiaoyu banfa yijian shu," 3.

71. Ninohei, *Nihon gyogyō kindai shi*, 58, 60–61, 93–97, 102; Okamoto Nobuo, *Kindai gyogyō hattatsu shi*, 257.

72. Okamoto Nobuo, *Kindai gyogyō hattatsu shi*, 111–12.

73. Zhang Zhuzun, "Fazhan Zhejiang shuichan jiaoyu banfa yijian shu," 4.

74. The original regulations are reprinted in Zhongguo dier lishi dang'anguan, *Zhonghua minguo shi dang'an ziliao huibian*, vol. 1, 3: 576.

75. Li Shihao and Qu Ruoqian, *Zhongguo yuye shi*, 19.

76. Ibid., 686.

77. Ibid.

78. "Haizhou yuye jishu chuanxisuo wei baosong Qingdao yuye diaocha baogaoshu cheng gao," April 7, 1922, in Zhonguo dier lishi dang'anguan, *Zhonghua minguo shi dang'an ziliao huibian*, vol. 1, 3: 705.

79. Ibid., 695, 701–7. See also "Wang Wentai mibao Ri ren jingying Qingdao yuye you ai woguo haiquan yuli ji jieshou shi sheshi jielüe gao," May 25, 1922, in ibid., 716–17.

80. "Wang Wentai mibao Ri ren jingying Qingdao yuye you ai woguo haiquan yuli ji jieshou shi sheshi jielüe gao," May 25, 1922, in Zhonguo dier lishi dang'anguan, *Zhonghua minguo shi dang'an ziliao huibian*, vol. 1, 3: 716–17.

81. "Wang Wentai chen song Haizhou yuye jihuashu cheng gao," February 16, 1920, in Zhonguo dier lishi dang'anguan, *Zhonghua minguo shi dang'an ziliao huibian*, vol. 1, 3: 721.

82. "Haizhou yuye jishu chuanxisuo wei baosong Qingdao yuye diaocha baogaoshu cheng gao," April 7, 1922, Zhonguo dier lishi dang'anguan, *Zhonghua minguo shi dang'an ziliao huibian*, vol. 1, 3: 704–6. See also Li Shihao and Qu Ruoqian, *Zhongguo yuye shi*, 199–200.

83. "Wang Wentai mibao Ri ren jingying Qingdao yuye you ai woguo haiquan yuli ji jieshou shi sheshi jielüe gao," in Zhonguo dier lishi dang'anguan, *Zhonghua minguo shi dang'an ziliao huibian*, vol. 1, 3: 716.

84. Shindo, "A Statistical Account of the Japanese Trawl Fishery in the East China and the Yellow Seas After the War II [*sic*]," 2–3.

85. "Haizhou yuye jishu chuanxisuo wei baosong Qingdao yuye diaocha baogaoshu cheng gao," April 7, 1922, in Zhonguo dier lishi danganguan, *Zhonghua minguo shi dang'an ziliao huibian*, vol. 1, 3: 692; "Wang Wentai mibao Ri ren jingying Qingdao yuye you ai woguo haiquan yuli ji jieshou shi sheshi jielüe gao," in Zhonguo dier lishi dang'anguan, *Zhonghua minguo shi dang'an ziliao huibian*, vol. 1, 3: 715.

86. "Wang Wentai mibao Ri ren jingying Qingdao yuye you ai woguo haiquan yuli ji jieshou shi sheshi jielüe gao," in Zhonguo dier lishi dang'anguan, *Zhonghua minguo shi dang'an ziliao huibian*, vol. 1, 3: 715.

87. Ibid., 720–25.

88. Zhoushan shi dang'anguan, *"Shenbao" Zhoushan shiliao huibian*, 161.

89. This outline also included provisions for the reorganization of fishery instruction offices, allocating 23,000 *yuan* for the construction of a fishing trawler for the use of the Haizhou Fishery Training Institute. The Ministry of Agriculture and Commerce also planned to send students to Europe and North America to make up for the shortage of technical expertise needed for fishery reforms (*SB*, July 14, 1921: 11).

90. The Ministry of the Navy also had recently announced a desire to reform China's fishing industry, expressing its intention to send vessels to patrol fishing grounds and collect fishing taxes. These schemes emerged when fishing boats in Wenzhou and Taizhou requested the navy's protection after suffering attacks by pirates. The Ministry of Agriculture and Commerce objected, stating that reform of China's fishing industry, as well as the revenues that it generated, belonged under its jurisdiction (*SB*, July 14, 1921, 11). For additional articles on the Ministry of the Navy's fishing taxes, see *SB*, August 31, 1921, 11.

91. *SB*, July 14, 1921, 11.

92. *SB*, March 3, 1924, 15.

93. Sun Yat-sen, *The International Development of China*, 155–59. For discussion of Sun's *Industrial Plan*, see Kirby, "Engineering China," 138–39.

94. Jin Zhiquan, "Yu shichang zhi jianshe," 108. See also Hu Juntai, "Zhejiang shuichan zhi wojian," 9–10.

95. Wang Tang, "Duiyu zhengdun Zhongguo yuye zhi guanjian," *SB*, June 22, 1924 (supplement), 3.

96. See Yixin Chen, "The Guomindang's Approach to Rural Socioeconomic Problems."

97. Jin Zhao (1901–79) was born in Linhai county and graduated from the Zhejiang First Provincial Fishery School (Zhejiang shengli jiazhong shuichan xuexiao) in 1920. The following year he began studies at Hokkaido Imperial University's Department of Fisheries, graduating in 1925. He taught at the Zhejiang First Provincial Fishery School as well as the Jiangsu Provincial Fishery School before returning to Japan as a researcher at Tokyo Imperial University's fishery research laboratory in 1928 (Zhejiang sheng shuichanzhi bianzuan weiyuanhui, *Zhejiang sheng shuichanzhi*, 1015).

98. Jin Zhaohua, "Zhejiang shuichan jianshe wenti zhi jiantao," 3.

99. Ibid.

100. Ibid., 5–6.

101. Ibid., 7–8.

102. Ibid., 11.

103. Ibid., 12.

104. *SB*, July 14, 1921, 11. According to this report, the same-trade associations were founded by fish merchants, whose interests were not the same as those of fishermen. Hence, there was no way to make certain that the leaders of these associations had the mutual interests of their members in mind.

105. Zhongguo dier lishi dang'anguan, *Zhonghua minguo shi dang'an ziliao huibian*, vol. 3, 2: 725–32, 743–45, 750–53.

106. Ibid., vol. 5, 1, 7: 642–47.

107. Kirby, "China Unincorporated," 44.

108. Ibid.

109. Zhongguo dier lishi dang'anguan, *Zhonghua minguo shi dang'an ziliao huibian*, vol. 3, 2: 745–50.

110. On fishery legislation in Meiji Japan, see Howell, *Capitalism from Within*, 97–100; and Ninohei, *Meiji gyogyō kaitaku shi*, chap. 4.

111. Zhongguo dier lishi dang'anguan, *Zhonghua minguo shi dang'an ziliao huibian*, vol. 5, 1, 7: 637–42. See also Xie Zhenmin, *Zhonghua minguo lifa shi*, 593–94.

112. *DHZB*, November 1, 1934, 2.

113. Yao Yongping, "Gaijin Zhejiang dahuangyu yuye ji zhizaoye zhi yijian," 13–14.

114. Wang Ningshi, *Zhejiang yanhai ge xian yu yan gaikuang*, 17.

115. Wenling xian xuzhi gao bianzuan weiyuanhui, *Wenling xian xuzhi gao*, 37. For controversies over the registration fees that the Rules for the Implementation of the Temporary Fishing Association Regulations allowed county governments to collect from fishing boats and fish merchants, see *SB*, August 12, 1926, 9.

116. Wenling xian xuzhi gao bianzuan weiyuanhui, *Wenling xian xuzhi gao*, 37. See also "Wenling yuhui mingce," 1936, IMH 17-27 125-1.

117. Wenling xian xuzhi gao bianzuan weiyuanhui, *Wenling xian xuzhi gao*, 317. Chen Zhongxiu, who was born in Songmen in 1897, graduated from the Zhejiang Provincial Normal School. During the early years of the Republican period, Chen promoted the reform of popular education in his native place, contributed to local water-control projects, and directed local militia forces. See "Wenling yuhui mingce," 1936, IMH 17-27 125-1.

118. Wang Ningshi, *Zhejiang yanhai ge xian yu yan gaikuang*, 21.

119. Bao Heng, who was commonly referred to by his style name Zuoren, was also manager of Ningbo's Taizhou Native-Place Lodge (Lü Yong Taizhou gongsuo) and director of the Nanding Lodge in Gaoting and the Zhe-Tai Fishing Lodge in Shengshan; see "Wenling yuhui mingce," 1936, IMH 17-27 125-1; Wenling xian xuzhi gao bianzuan weiyuanhui, *Wenling xian xuzhi gao*, 277; and Zhao Yizhong, "Zhoushan de yuye gongsuo," 68.

120. "Zhejiang Yin xian yuhui," 1932–34, IMH 17-27 127-2; Zhao Yizhong, "Zhoushan de yuye gongsuo," 70.

121. Zhang Shenzhi (1877–1952) also went by his given name Chuanbao. For biographical information, see Zhejiang sheng Yin xianzhi bianzuan weiyuanhui, *Yin xianzhi*; and Zhejiang sheng zhengxie wenshi zilao weiyuanhui, *Zhejiang jinxiandai renwu lu*, 179–80.

122. Koppes, "Efficiency, Equality, Esthetics," 234. For the influence of "wise-use utilitarians" in Australia during the early twentieth century, see Frawley, "Evolving Visions," 66.

123. Frawley, "Evolving Visions," 67; Koppes, "Efficiency, Equality, Esthetics," 234.

124. McEvoy, *The Fisherman's Problem*, 108.

125. Josephson, *Resources Under Regimes*, 5–6, 197–98.

126. This developmental ideology is reminiscent of James Scott's analysis of scientific forestry in modern Europe, but Scott's stark opposition between the distinct realms of society and the state does not coincide well with the Chinese experience; see Scott, *Seeing Like a State*, 21.

Chapter 4

1. "Zhongguo shiye yinhang Ningbo banshichu wei she Daishan zhi Hu hang han," April 18, 1933, SMA Q276-1-613. See also Zhoushan yuzhi bianxie zu, *Zhoushan yuzhi*, 22. Both the Dinghai branches of the Bank of China and the Commercial Bank of China opened offices in the port of Dongshajiao on Daishan Island in May 1934 (*DHZB*, May 21, 1934, 2–3; June 6, 1932, 1–2; and Tu Hengting, "Minguo shiqi Daishan jinrongye gaikuang," 15–16).

2. Tu Hengting, "Minguo shiqi Daishan jinrongye gaikuang," 14–16; Wang Zongpei, "Zhongguo yanhai zhi yumin jingji," 134–36.

3. "Zhongguo shiye yinhang Shanghai fenhang lianhang Shenjiamen banshichu guanyu yingye jingguo ji fangzhen baogao ji caiche deng shiyi lai han," January 4, 1935, SMA Q276-1-614.

4. "Zhongguo tongshang yinhang guanyu chazhang baogao ji qiankuan cunshou Daishan duihuanchu bing renshi waidiao cizhi jiaxin baozheng deng yu Dinghai banshichu wanglai wenshu," March 17, 1934, SMA 281-1-545. See also Shenjiamen zhenzhi bianzuan lingdao xiaozu, *Shenjiamen zhenzhi*, 362.

5. For Chinese merchants' negative attitudes toward taking out loans on security, see Linsun Cheng, *Banking in Modern China*, 146–53.

6. "Zhongguo shiye yinhang Shanghai fenhang lianhang Shenjiamen banshichu guanyu yingye jingguo ji fangzhen baogao ji caiche deng shiyi laihan," March 9, March 15, April 11, and May 3, 1935, SMA Q276-1-614.

7. Liu Jiting (1890–1942) was a native of Shenjiamen. After working as an employee at a native bank during his youth, Liu was director of the Bank of China's Shenjiamen and Dinghai offices from 1922 to 1935 (Zhoushan shi bowuguan, *Zhoushan lishi mingren pu*, 146–47; Zhoushan shi difangzhi bianzuan weiyuanhui, *Zhoushan shizhi*, 798).

8. Zhang Xiaogeng (1889–1939), a member of the Zhang lineage in Dinghai's Zhanmao village, moved to Shenjiamen as a child. After studying law in Japan for a time, he founded the Dinghai branch of the Commercial Bank of China in 1925. Zhang stepped down as the bank's director that same year when he was selected head of the Dinghai county assembly. Zhang also managed several other business enterprises related to the fishing industry. These included the Yongxin Refrigeration Company founded in Shanghai during the 1930s and a firm in Ningbo that traded in the pig's blood used to preserve fishing nets (Zhoushan shi bowuguan, *Zhoushan lishi mingren pu*, 133–34).

9. "Zhejiang Yin xian yuhui," 1932–34, IMH 17-27 127-2; Zhao Yizhong, "Zhoushan de yuye gongsuo," 70.

10. See Zhoushan yuzhi bianxie zu, *Zhoushan yuzhi*, 109–10.

11. Lin Shuyan and Huang Shubiao, *Zhejiang zhangwang yingxiang yulei fanzhi zhi yanjiu*.

12. Ibid., 27; Yao Yongping, "Gaijin Zhejiang dahuangyu yuye ji zhizaoye zhi yijian," 16.

13. As Vaclav Smil (*China's Past, China's Future*, 109–20) has shown, synthetic fertilizer was not widely used in China until as late as the 1970s.

14. *Kōsō shō Sekkō shō suisangyō chōsa hōkoku*, 30.

15. Lin Shuyan and Huang Shubiao, *Zhejiang zhangwang yingxiang yulei fanzhi zhi yanjiu*, 26.

16. Iversen, *Living Marine Resources*, 247; Jennings et al., *Marine Fisheries Ecology*, 145.

17. *Minguo Xiangshan xianzhi* 13.28b.

18. *Taizhou fuzhi* 60.63a.

19. *Daishan zhenzhi* 20.9ab.

20. Precipitation levels over southeast China and the Lower Yangzi region underwent a relative decline during the 1920s. The resulting reduction of runoff from the Yangzi River may have adversely affected

primary productivity and reproduction of fish stocks. See Chang and King, "Centennial Climate Changes in the Yangtze River Delta," 99; Yang et al., "Trends in Annual Discharge from the Yangtze River to the Sea," 828–31; and Zhang Jiacheng and Lin Zhiguang, *Climate of China*, 297–98.

21. Bolster, "Putting the Ocean in Atlantic History."

22. Jiang Bin and Jin Tao, *Donghai daoyu wenhua yu minsu*, 95. Again, it is hard to tell if human agency or fluctuating ocean conditions led to this shift.

23. *DHZB*, May 11, 1934, 1–2. See also Jin Jifu, "Zhejiang Tai shu shui-chan gaikuang," 25. Unfortunately, records from the Jiangsu-Zhejiang Fishing Company necessary to verify the accuracy of these assertions do not exist.

24. Guo Zhenmin, *Shengsi yuye shihua*, 99; Jin Jifu, "Zhejiang Tai shu shuichan gaikuang," 4.

25. Cheng Tiyun, "Jiangsu waihai shan dao zhi," 6; Guo Zhenmin, *Shengsi yuye shihua*, 31; Jin Jifu, "Zhejiang Tai shu shuichan gaikuang," 24; Zhao Yizhong, "Zhoushan yuye fazhan shi chutan," 109–10; Zhoushan yuzhi bianxie zu, *Zhoushan yuzhi*, 24–25, 57–58, 65.

26. Jie Tianhai, "Caiyuan yuhang xingshuai ji," 29.

27. Guo Zhenmin, *Shengsi yuye shihua*, 165–67.

28. At least four of the seven largest fish brokerages on Shengshan Island during the 1930s and 1940s were run by merchants who had originally come from Daishan or Shenjiamen (ibid., 170–74).

29. Kemp, ed., *The Oxford Companion to Ships and the Sea*, 886.

30. Kibesaki, "Fundamental Studies on Structure and Effective Management of the Demersal Fish Resources in the East China and the Yellow Sea," 74–75; Okamoto Nobuo, *Kindai gyogyō hattatsu shi*, 166–70; Shindo, "A Statistical Account of the Japanese Trawl Fishery in the East China and the Yellow Seas After the War II [sic]," 2–3; Takenobu, *The Japan Year Book*, 513.

31. Okamoto Nobuo, *Kindai gyogyō hattatsu shi*, 178–79.

32. Shindo, "A Statistical Account of the Japanese Trawl Fishery in the East China and the Yellow Seas After the War II [sic]," 3–4.

33. Ibid., 5–6.

34. *Dinghai xianzhi* 5.6ab.

35. "Shou Shanghai zongshanghui dian," May 8, 1925, in *Zhong-Ri guanxi shiliao*, 486; "Shou zhi zhengfu mishuting han," May 11, 1925, in

ibid., 490; "Shou Nongshangbu zi," May 15, 1925, in ibid., 492; "Haizhou yuye shiyanchang changzhang cheng," April 28, 1925, in ibid., 493.

36. "Shou Jiangsu shengzhang (Zheng Qian) daidian," May 16, 1925, in *Zhong-Ri guanxi shiliao*, 494.

37. Gerth, *China Made*.

38. "Shou Shanghai dianju Xu Shaoqing daidian," May 15, 1925, in *Zhong-Ri guanxi shiliao,* 493.

39. "Shou Nongshangbu zi," May 15, 1925, in *Zhong-Ri guanxi shiliao*, 492.

40. On the Zhenhai Fang lineage's business activities, see Li Jian, *Shanghai de Ningboren*, 74; and Wang Renze, "Lü Hu mingren Fang Jiaobo," 100.

41. "Guomindang zhongzhiwei guanyu Jiang Zhe Hu yuye zhuangkuang han," December 24, 1932, in Zhongguo dier lishi dang'anguan, *Zhonghua minguo shi dang'an ziliao huibian*, vol. 5, 1, 7: 656.

42. Zhejiang sheng Yin xianzhi bianzuan weiyuanhui, *Yin xianzhi*, 2079–80; Zhejiang sheng zhengxie wenshi zilao weiyuanhui, *Zhejiang jinxiandai renwu lu*, 179–80.

43. "Fa Riben Fangze gongshi (Qianji) han," May 20, 1925, in *Zhong-Ri guanxi shiliao*, 497–98.

44. "Fa Nongshangbu gonghan," June 2, 1925, in *Zhong-Ri guanxi shiliao*, 510; "Shou Jiangsu jiaosheshu daidian," April 6, 1926, in ibid., 563; "Fa Haijun, Lujun, Jiaotong, Nongshangbu han," April 26, 1926, in ibid., 565; "Fa Jiangsu shengzhang (Chen Taoyi) han," April 27, 1926, in ibid., 566; "Fa Shandong shengzhang (Zhang Zongchang) han," October 21, 1926, in ibid., 611.

45. Scandinavian countries generally claimed four miles, whereas other countries such as France, Italy, Russia, and Spain claimed jurisdiction over any distance necessary to control fishing and smuggling; see Churchhill and Lowe, *The Law of the Sea*, 74, 77–79.

46. Reeves, "The Codification of the Law of Territorial Waters," 491. According to the legal scholar Tachi Sakutarō (*Heiji kokusaihō ron*, 304–5), Japan first decided on three miles as the limit of its territorial waters in 1871. Japan's consistent adherence to this principle during the early twentieth century is apparent in the cases cited in "The Law of Territorial Waters," 260, 340–341.

47. Churchhill and Lowe, *The Law of the Sea*, 14–15; Dhokalia, *The Codification of Public International Law*, 124–25. During the final meeting of the Territorial Waters Committee, twenty states supported territo-

rial seas of three miles, twelve argued for six miles, and four Scandinavian states sought recognition for their established four-mile claims. To make matters even more complicated, many countries favored the three-mile limit on the condition of the recognition of a contiguous zone of jurisdiction beyond the territorial sea.

48. Churchhill and Lowe, *The Law of the Sea*, 79.

49. Reeves, "The Codification of the Law of Territorial Waters," 492–93; "Report of the Second Committee," 254–55.

50. "Fa Shandong shengzhang (Zhang Zongchang) han," October 21, 1926, in *Zhong-Ri guanxi shiliao*, 616. In July 1912 the Ministry of the Navy originally recommended that China set its territorial waters at ten nautical miles from shore; no action was taken on the issue until the 1920s; see "Fa shuiwuchu zi," January 15, 1918, in ibid., 6–7.

51. "Fa ti guowu huiyi an," attached to "Fa zhi zhengfu mishuchu han," May 25, 1925, in *Zhong-Ri guanxi shiliao*, 504. At the Pacific Conference held in Honolulu during July 1924, however, the Chinese representative Ye Keliang, then serving as China's general consul in San Francisco, argued that international law allowed coastal countries to maintain fishing rights on the high seas for their own people's use. Ye also maintained that unregulated fishing was bringing about a gradual depletion of the Pacific Ocean's marine resources. Hence, he argued that all countries bordering on the Pacific Ocean should establish common legislation to prevent violation of the fishing territories of neighboring countries ("Fa Shandong shengzhang [Zhang Zongchang] han," October 21, 1926, in ibid., 612–13).

52. *SB*, April 27, 1926, 13.

53. The Zhejiang-Jiangsu War was waged between the warlord who controlled Jiangsu and the one in control of Zhejiang. Zhejiang's warlord held Shanghai at the time. For a detailed narrative of this conflict, see Waldron, *From War to Nationalism*, chap. 4.

54. Ibid., 125–27.

55. *SB*, July 18, 1924, 14; April 27, 1925, 14.

56. *SB*, February 27, 1924, 3.

57. *SB*, June 10, 1924, 14; July 5, 1924, 14.

58. *SB*, July 24, 1925, 11.

59. *SB*, June 10, 1924, 14; July 5, 1924, 14.

60. According to Wang's proclamation, the Fishery Defense Bureau would facilitate official management of Jiangsu's fisheries by registering fishing boats and compiling population data for coastal fishing centers.

The bureau would also assert tighter control over revenues by reorganizing fishing militias and eliminating all informal fees collected from fishing enterprises (*SB*, June 10, 1924, 14; July 24, 1925, 11).

61. *SB*, July 24, 1925, 11.

62. *SB*, May 9, 1926, 9; July 11, 1926, 15; August 1, 1926, 13.

63. *SB*, August 2, 1926, 13.

64. *SB*, June 10, 1926, 14; July 6, 1926, 15; August 11, 1926, 13; August 14, 1926, 15.

65. Li Shihao and Qu Ruoqian, *Zhongguo yuye shi*, 45.

66. Ibid., 45–50. See also *SB*, August 8, 1927, 9; November 20, 1927, 9.

67. *SB*, April 1, 1928, 8.

68. *SB*, August 8, 1927, 9; December 6, 1927, 15. See also Li Shihao and Qu Ruoqian, *Zhongguo yuye shi*, 50–51.

69. "Cheng Nongkuangbu," August 18, 1928, in *Zhonghua minguo shuichan xuehui huibao* (1934): 15–16, SMA Y4-1-225. For the Fishery Study Association's petitions for the inclusion of a fishery management agency in the Nationalist government's newly organized bureaucracy, see "Zhonghua minguo shuichan xuehui cheng guomin zhengfu Jianshe weiyuanhui ji Nongkuangbu wen," August 1927, in *Zhonghua minguo shuichan xuehui huibao*, 14–15, SMA Y4-1-225.

70. The conference assembled representatives from the Nationalist government, Zhejiang, Jiangsu, and the Shanghai municipal government (Zhejiang sheng, Jiansheting, *Liang nian lai zhi Zhejiang jianshe gaikuang*, 4: 13).

71. To fund these projects, along with additional research and education institutes, fishery experts argued that the government should set aside "fishery reconstruction fees" (*yuye jianshe fei*) obtained from fishing tax revenues (*SB*, September 30, 1928, 16).

72. *SB*, October 5, 1928, 16; October 6, 1928, 16.

73. "Zhedong yumin daibiao Shi Renhang deng daidian," May 1, 1930, attached to "Zhejiang sheng zhengfu zi," November 25, 1930, IMH 17-27 6-1.

74. "Riben yulun qin ru Jiang-Zhe yangmian," April 8, 1931; "Zhuang Songfu cheng," April 4, 1931, IMH 17-27 7-1. Chiang Kai-shek reportedly viewed his fellow Fenghua-native Zhuang Songfu as one of his elders; so he was able to speak to Chiang in a direct manner (Mao Yihu, "Zhuang Songfu de yisheng," 37).

75. "Zhedong yumin daibiao Shi Renhang deng daidian," n.d., IMH 17-27 6-1.

76. *DFZZ*, September 10, 1930, 8–10; February 25, 1930, 3–5. See also Tung, *China and Some Phases of International Law*, 9.

77. Huang Zhenshi, "Jiu Shanghai de yushi," 249–55; Li Jian, *Shanghai de Ningboren*, 369–70.

78. "Gongshangbu zi," November 25, 1930, IMH 17-27 6-1.

79. "Dian qing yanhai ge sheng dangbu shi dangbu dian zhongyang chi wu yue yi ri Ri yulun jinzhi jinkou yuan yi you," December 15, 1928, in *Zhonghua minguo shuichan xuehui huibao* (1934): 19, SMA Y4-1-225.

80. The event was organized by fishery specialists and Nationalist Party branches in Jiangsu to rouse support for reform and development of Jiangsu's fisheries (*Shuichan xuebao* 1, no. 1 [1931]: 122–23). See also Li Shihao and Qu Ruoqian, *Zhongguo yuye shi*, 88–90. On Kong Xiangxi's attempts to obtain official financial support to fund programs for China's industrial development, see Jordan, *Chinese Boycotts Versus Japanese Bombs*, 55, 57–59, 69, 331.

81. "Xingzhengyuan zhiling," February 27, 1931; "Xingzhengyuan xunling," February 18, 1931; "Guiding linghai jiexian caoan," April 1931, in "Chonghua woguo linghai jiexian ji jisi wenti quan an chao," AH 127-1393. Copies of documents related to China's territorial waters are also found in AH 062-838.

82. "Xingzhengyuan xunling," April 28, 1931, AH 127-1393. See also *Shuichan xuebao*, 1, no. 1 (1931): 127.

83. "Xingzhengyuan xunling," April 28, 1931, AH 127-1393.

84. "Ri daiban yuan han yiwen," February 17, 1931, IMH 17-27 6-1.

85. "Yi Zhongguo haigang wei genjudi zhi Riben tuowangchuan congshi yuanyang yuye zhi xianzhuang," IMH 17-27 7-1. For later Chinese reports confirming this contention, see "Huyu banshichu cheng," November 22, 1934, IMH 17-27 6-1.

86. "Shiyebuzhang Kong Xiangxi ti'an yi," February 19, 1931, IMH 17-27 6-1; Li Shihao and Qu Ruoquan, *Zhongguo yuye shi,* 206.

87. "Shiyebuzhang Kong Xiangxi ti'an yi," February 19, 1931, IMH 17-27 6-1; Li Shihao and Qu Ruoqian, *Zhongguo yuye shi,* 207–8.

88. "Xingzhengyuan ling," February 26, 1931, IMH 17-27 6-1.

89. Li Shihao and Qu Ruoqian, *Zhongguo yuye shi,* 53–54, 209.

90. "Guomin zhengfu wenguanchu wei yushang hu yang shi yu Xingzhengyuan wanglai han," April 9–12, 1931, in Zhongguo dier lishi dang'anguan, *Zhonghua minguo shi dang'an ziliao huibian,* vol. 5, 1, 7: 649–52.

91. "Riben gongshiguan jielüe," March 13, 1931, AH 172-1 3174. See also "Yi Zhongguo haigang wei genjudi zhi Riben yewangchuan congshi yuanyang yuye zhi xianzhuang," IMH 17-27 7-1.

92. "Caizhengbu mizi," April 16, 1931; "Waijiaobu mizi," April 4, 1931; "Waijiaobu mizi," April 24, 1931, IMH 17-27 7-1.

93. "Caizhengbu daidian," March 2, 1931; "Caizhengbu zi," April 27, 1931, IMH 17-27 7-1.

94. When Japan finally agreed to tariff autonomy in 1930, it did so on the conditions that 5 million taels from China's customs receipts would be set aside to repay foreign debts and that reduced customs duties would be charged for certain categories of Japanese imports (Iriye, *China and Japan in the Global Setting*, 56–59). Imported Japanese fishery products were among the goods subject to lowered rates (Li Shihao and Qu Ruoqian, *Zhongguo yuye shi*, 182–83).

95. Iriye, *After Imperialism*, 289–93.

96. Coble, *Facing Japan*, 22; Iriye, *China and Japan in the Global Setting*, 51.

97. "Caizhengbu daidian," April 27, 1931, IMH 17-27 7-1.

98. On this "factional" struggle between the Ministry of Industy and the Ministry of Finance, see Jordan, *Chinese Boycotts Versus Japanese Bombs*, 53–54.

99. "Waijiaobu zi," April 30, 1931, IMH 17-27 7-1.

100. "Shiyebu zi," May 6, 1931, IMH 17-27 7-1; "Liu yue wuqian shi shi zai Waijiaobu kai hui," June 20, 1931, AH 172-1 3174.

101. "Caizhengbu daidian," July 20, 1931; "Caizhengbu daidian," August 7, 1931, IMH 17-27 6-2.

102. Memo from the Department of Fisheries and Animal Husbandry, August 11, 1931, IMH 17-27 6-2.

103. "Yumusi cheng," August 3, 1931, IMH 17-27 6-2.

104. "Shanghai shi zhengfu zi," December 1, 1934, IMH 17-27 6-1. See also Li Shihao and Qu Ruoqian, *Zhongguo yuye shi*, 109.

105. On the Shanghai Incident, see Coble, *Facing Japan*, 39–50; and Jordan, *Chinese Boycotts Versus Japanese Bombs*, chap. 18.

106. "Shanghai shi zhengfu zi," December 1, 1934, IMH 17-27 6-1.

107. "Jiangsu sheng zhengfu zi," November 14, 1934, IMH 17-27 6-1.

108. Wu Xingya, *Minguo ershiyi nian Shanghai shi yulun huigu*, 31, 33–34.

109. Lin Shuyan and Huang Shubiao, *Zhejiang zhangwang yingxiang yulei fanzhi zhi yanjiu*, 29–44.

110. Cao Erhui, "Su sheng waihai jiushi yuye zhi," 9; Wu Xingya, *Minguo ershiyi nian Shanghai shi yulun huigu*, 34. Eighteen Chinese-managed fishing companies operated a total of 31 trawlers out of Shanghai in 1934 (*Shanghai shi nianjian*, 17: 42–44).

111. Zhang Baoshu, *Zhongguo yuye*, 194.

112. Okamoto Nobuo, *Kindai gyogyō hattatsu shi*, 375–76.

113. Yao Yongping, a native of Jiangsu, graduated from the Jiangsu Provincial Fishery School and studied in Japan before becoming an instructor at the Zhejiang Provincial Fishery School ("Huiyuan lu," in *Zhonghua minguo shuichan xuehui huibao* [1934], 107).

114. Yao Yongping, "Daishan shuichan zhi diaocha," 14.

115. Ibid.

116. *DHZB*, June 6, 1934, 2.

117. *DHZB*, October 16, 1934, 2.

118. *DHZB*, July 2, 1936, 2.

Chapter 5

1. Guo Zhenmin, *Shengsi yuye shihua*, 91–92; Zhoushan yuzhi bianxie zu, *Zhoushan yuzhi*, 59.

2. Pettus, "Cuttle-fish Trade at Ningpo," 550.

3. Ibid., 551.

4. Worcester, *The Floating Population in China*, 24.

5. *Dinghai xianzhi* 5.2a–3a; Cao Erhui, "Su sheng waihai jiushi yuye zhi," 2–6.

6. Members of the Republic of China Fishery Studies Association advised the Jiangsu provincial government on the formation of its fishery policies and took part in the Jiangsu Province Agricultural Policy Conference held in October 1928 ("Huiwu jingguo gaikuang," 2, in *Zhonghua minguo shuichan xuehui huibao* [1934], SMA Y4-1-225).

7. Okano, "Shina shin seifu no suisan seisaku," 15–19.

8. "Quanguo shuichan shuxue jiguan gaikuang," in *Zhonghua minguo shuichan xuehui huibao* (1934): 81–82, SMA Y4-1-225. See also Guo Zhenmin, *Shengsi yuye shihua*, 317–21. A native of Jiangsu, Zhang Zhuzun had graduated from the Jiangsu Provincial Fishery School and studied abroad in Japan before becoming the director of the Zhejiang Provincial Fishery School. See "Huiyuan lu," in *Zhonghua minguo shuichan xuehui huibao* (1934), 107, SMA Y4-1-225.

9. Wang Liansheng was a native of Jiangsu who had graduated from the Zhili Provincial Fishery School and worked as a technician at the Haizhou Fishery Experiment Station during the 1920s. See "Huiyuan lu," in *Zhonghua minguo shuichan xuehui huibao* (1934), 101, SMA Y4-1-225.

10. "Quanguo shuichan shuxue jiguan gaikuang," in *Zhonghua minguo shuichan xuehui huibao* (1934): 84–85, SMA Y4-1-225.

11. Guo Zhenmin, *Shengsi yuye shihua*, 308–9.

12. Ibid., 7.

13. Lu Yanghao, "Pi Zhe sheng 'Ningbo,' 'Fenghua,' 'Dinghai' Hu tong-xianghui qing hua Shengsi liedao gui Zhe shuo," 2.

14. Taizhou cage fishers' catches were generally higher than those of cage fishers from Wenzhou since Taizhou natives monopolized the best fishing spots and used cages made from bamboo that attracted cuttlefish by reflecting light (*Zhongguo shiyezhi, Zhejiang sheng*, part 5, section 2, 16–17).

15. "Zhoushan qundao zhi yuchang," 12.

16. Zhejiang sheng, Shuiliju, *Zhejiang sheng shuiliju zongbaogao*, 1: 104–6, 126. On the 1931 floods in Zhejiang, see Zhou Kebao and Lu Yonglong, "Ershi nian Zhejiang sheng zhi shuizai."

17. Yao Huanzhou, "Zhoushan qundao wuzei zhi shengxi ji wangbu yu longbu zhi deshi," 41–42.

18. "Jiang Zhe yanhai zhi moyu yuchang ji yumin shenghuo," 83; Yao Huanzhou, "Zhoushan qundao wuzei zhi shengxi ji wangbu yu longbu zhi deshi," 40.

19. *Zhongguo nongye baikequanshu*, vol. 11, 306–7.

20. Yao Yongping, "Gaijin Zhejiang dahuangyu yuye ji zhizaoye zhi yijian," 16.

21. Gary Libecap (*Contracting for Property Rights*, 87) notes that this is true of most "private group controls" on fishery resources.

22. Cao Erhui, "Su sheng waihai jiushi yuye zhi," 7.

23. Ibid.

24. Guo Zhenmin, *Shengsi yuye shihua*, 117.

25. Ibid., 260. "Jiangsu sheng Jiansheting tingzhang He Yushu cheng," May 4, 1932, IMH 17-27 11-1.

26. "Zhejiang Huangyan yumin daibiao Zhang Lezhai deng suyuan-shu," September 2, 1931, IMH 17-27 135-6.

27. James M. Acheson (*Capturing the Commons*, 227) stresses the complex and unpredictable nature of fishery production.

28. Zhoushan yuzhi bianxie zu, *Zhoushan yuzhi*, 65.

29. Imperial Maritime Customs, *Returns of Trade and Trade Reports*, 322.

30. Qiu Yongsong et al., "Runoff- and Monsoon-Driven Variability of Fish Production in East China Seas," 31.

31. On the high level of rainfall in central China during 1931, see Zhang Jiacheng and Lin Zhiguang, *Climate of China*, 290.

32. The statute of limitations for appealing the Jiangsu-Zhejiang Fishery Management Bureau's decision had expired, and complaints regarding Jiangsu's and Zhejiang's decisions had to go through the provinces before being brought to the central government (memo from Su-yuan shenli weiyuanhui to Yumusi, October 15, 1931; "Shiyebu pi Zhe-jiang Huangyan xian yumin Zhang Lezhai deng," October 20, 1931, IMH 17-27 135-6).

33. Fishery Office memos, October 6 and 15, 1931, IMH 17-27 135-6.

34. Zhejiang Reconstruction Office to Ministry of Industry, May 4, 1932, IMH 17-27 11-1.

35. Kuhn, "Local Self-Government Under the Republic," 295.

36. Wenling xian xuzhi gao bianzuan weiyuanhui, *Wenling xian xuzhi gao*, 280.

37. "Wenling yuhui mingce," IMH 17-27 125-1.

38. *CMB*, January 1, 1933, 19; Lu Yanghao, "Pi Zhe sheng 'Ningbo,' 'Fenghua,' 'Dinghai' Hu tongxianghui qing hua Shengsi liedao gui Zhe shuo," 2; Ping, "Shatian xian ken chuyan," 9–11. On abuses and corruption in the polder-land business, see Zhu Fucheng, *Jiangsu shatian zhi yanjiu*, 36081–100.

39. *CMB*, January 1, 1933, 17; Du Haogeng, "Moyu bulong wenti de jian-tao," 10.

40. "Chongming xian yuhui Shengshan fenhui daidian," April 23, 1932, IMH 17-27 11-1; Li Shihao and Qu Ruoqian, *Zhongguo yuye shi*, 54–55.

41. "Jiang-Zhe qu yuye guanliju juzhang Han Yougang cheng," April 1, 1932, IMH 17-27 11-1.

42. "Chongming xian yuhui Shengshan fenhui daidian," April 23, 1932, IMH 17-27 11-1.

43. "Ningbo lü Jing tongxianghui cheng," May 3, 1932, IMH 17-27 11-1.

44. "Ningbo lü Hu tongxianghui daidian," May 16, 1932, IMH 17-27 11-1; "Lü Hu Taizhou liu yi tongxianghui daidian," June 13, 1932, IMH 17-27 11-2.

45. "Shiyebu zhiling," April 13, 1932, IMH 17-27 11-1.

46. See the two petitions from He Yushu to the Ministry of Industry dated April 30, 1932, IMH 17-27 11-1.

47. Li Shixiang memo, May 9, 1932, IMH 17-27 11-1.

48. "Shiyebu ling," May 13, 1932, IMH 17-27 11-1.

49. "Zhejiang sheng Jiansheting tingzhang Zeng Yangfu cheng," May 4, 1932; "Shiyebu zhiling," May 14, 1932, IMH 17-27 11-1.

50. "Jiangsu sheng Shiyeting cheng," April 26, 1932, IMH 17-27 120-5.

51. "Jiangsu sheng Shiyeting xunling di 1989 hao," June 1932, attached to "Yumin Mo Fuchang deng cheng," February 22, 1933, IMH 17-27 11-2; and Lu Yanghao, "Pi Zhe sheng 'Ningbo,' 'Fenghua,' 'Dinghai' Hu tong-xianghui qing hua Shengsi liedao gui Zhe shuo," 3. For Ge Liquan's lodge position, see Zhongguo dier lishi dang'anguan, *Zhonghua minguo shi dang-an ziliao huibian*, vol. 5, 1–7, 651, 656.

52. "Chuangban sili Shengshan yumin xiaoxue," *CMB*, April 4, 1934, 1.

53. "Jiangsu shengli yuye shiyanchang fushe yuye zhidaosuo qishi," April 29, 1932, IMH 17-27 11-2.

54. "Shiyebu daidian," June 18, 1932, IMH 17-27 11-2; Guo Zhenmin, *Shengsi yuye shihua*, 9, 261; Liu Tongshan and Xu Jibo, "Zhongguo yanhai yuye yu yumin shenghuo," 96–97.

55. "Jiangsu, Zhejiang liang sheng huafen yanhai ge daoyu jiexian di yi ci huiyi jilu," July 5, 1932, IMH 17-27 28-3.

56. Qu Ruoquan, "Jiang Zhe zhengyi zhong zhi Shengsi huazhi wen-ti," 1–2.

57. A native of Hunan, He Huiyu had graduated from the Tokyo Fishery Institute before serving as a technical advisor for the Ministry of Industry and the Zhejiang provincial government ("Huiyuan lu," in *Zhonghua minguo shuichan xuehui huibao* [1934] 12, SMA Y4-1-225).

58. "Jiangsu, Zhejiang liang sheng huafen yanhai ge daoyu jiexian di yi ci huiyi jilu," July 5, 1932, IMH 17-27 28-3.

59. "Shiye, Neizhengbu zi Jiangsu, Zhejiang sheng zhengfu," July 11, 1932, IMH 17-27 28-3.

60. Qu Ruoquan, "Jiang Zhe zhengyi zhong zhi Shengsi huazhi wenti," 2.

61. *CMB*, February 2, 1933, 2–3; February 21, 1933, 2.

62. Their claim was that, according to the procedures spelled out in the Fishery Law, the Ministry of Industry had violated the Fishery Management Bureau's administrative authority by annulling its order lifting the prohibition on the basis of Jiangsu's request. Cage fishers' representatives also argued that the Ministry of Industry violated China's Criminal

Code by denying cage fishers their legal right to carry out their profession ("Dailiren lüshi Cao Lin cheng," June 14, 1932, IMH 17-27 11-2; see also Guo Zhenmin, *Shengsi yuye shihua*, 262).

63. "Xingzhengyuan juedingshu," June 23, 1932, IMH 17-27 11-2.
64. "Yumin Mo Fuchang deng cheng," February 1933, IMH 17-27 11-2.
65. "Shiyebu pi," March 17, 1933, IMH 17-27 11-2.
66. "Zhejiang sheng zhengfu zi," August 2, 1932, IMH 17-27 12-1.
67. "Shiyebu cheng Xingzhengyuan," December 18, 1932, IMH 17-27 12-1. The Administrative Yuan approved this proposal in early March ("Xingzhengyuan xunling," March 4, 1933, IMH 17-27 12-2).
68. "Shiyebu ling Jiang-Zhe qu yuye gaijin weiyuanhui," February 24, 1933, IMH 17-27 12-2. Listings of the membership of the Fishery Improvement Commission can be found in "Weiyuan mingdan lüli biao ji shidao tuanti mingdan," February 15, 1933, IMH 17-27 52-1; and Li Shihao and Qu Ruoqian, *Zhongguo yuye shi*, 56. On Du Yuesheng and his fellow gangster Huang Jinrong's connections with the Shanghai fish business, see Martin, *The Shanghai Green Gang*, 208–9.
69. Hou Chaohai, "Shiyebu Jiang-Zhe qu yuye gaijin weiyuanhui chengli shi yuzheng zhi sheshi tan," 32.
70. Li Shihao and Qu Ruoqian, *Zhongguo yuye shi*, 57–58.
71. Huang Zhenshi, "Jiu Shanghai de yushi," 230.
72. Li Shihao and Qu Ruoqian, *Zhongguo yuye shi*, 58, 61.
73. Huang Zhenshi, "Jiu Shanghai de yushi," 231; Li Shihao and Qu Ruoqian, *Zhongguo yuye shi*, 58–59.
74. "Shiyebu Jiang-Zhe qu yuye gaijin weiyuanhui zhuxi weiyuan Chen Gongbo cheng," October 28, 1933, IMH 17-27 12-2; Du Haogeng, "Moyu bulong wenti de jiantao," 11.
75. Du Haogeng, "Moyu bulong wenti de jiantao," 11–12.
76. Ibid., 12. See also "Zhe shuichanchang niding moyu yuye qudi banfa," *SCYK* 3, no. 3–4 (1936): 108–10.
77. "Zhejiang shuichanchang niding moyu yuye qudi banfa," *SCYK* 3, no. 3–4 (1936): 109.
78. Ibid; Yao Huanzhou, "Zhoushan qundao wuzei zhi shengxi ji wangbu yu longbu zhi deshi," 53.
79. "Zhejiang shuichanchang niding moyu yuye qudi banfa," *SCYK* 3, no. 3–4 (1936): 108–9.
80. For an excellent overview of international scientific debates on the effects of fishing, see Smith, *Scaling Fisheries*.

81. Yao Huanzhou, "Zhoushan qundao wuzei zhi shengxi ji wangbu yu longbu zhi deshi," 61; Zhejiang shengli shuichan shiyanchang, "Sheng-shan moyu fanzhi shiyan baogao," 31.

82. On the concept of "lead agencies," see Young, *The Institutional Dimensions of Environmental Change*, 71, 127–28.

83. "Jiangsu sheng Jiansheting cheng," February 2, 1935; "Huyu ban-shichu cheng," April 3, 1935; "Zhejiang sheng zhengfu cheng," April 10, 1935, IMH 17-27 12-5.

84. "Zhiling benbu Huyu banshichu," May 24, 1935, IMH 17-27 12-5.

85. See Du Yuesheng's memo, March 13, 1934; "Hanfu Du Yuesheng xiansheng han gao," March 30, 1934; "Tai, Wen longbu yumin daibiao Chen Shan deng cheng," March 12, 1934; "Pi Tai Wen longbu yumin daibiao Chen Shan deng," March 19, 1934, IMH 17-27 11-2.

86. Guo Zhenmin, *Shengsi yuye shihua*, 263.

87. Changes in rules pertaining to the allocation and use of common-pool resources result from inherently political negotiations among what Gary Libecap (*Contracting for Property Rights*, 4, 26–28) terms "contract-ing parties." In his definition, these parties consist of private claimants, political actors, and bureaucratic organs.

Chapter 6

1. The business tax originated with fiscal reforms instituted during the Nationalist government's Second Economic Conference of 1934, which attempted to unify tax collection in county-level offices under the authority of provincial governments. By creating a direct line of author-ity between provincial revenue departments and county-level bureaus, this new fiscal policy was intended to increase bureaucratic control over business profits and consolidate tax collection by removing intermediar-ies at the local level. See Mann, *Local Merchants and the Chinese Bureau-cracy*, 169.

2. "Chongming xian yingyeshui zhengshouchu cheng Caizhengting wen," January 1935; "Chongming xian zhengfu, Jiangsu sheng Caizheng-ting zhishu Chongming yingyeshui zhengshouchu bugao," March 1935, IMH 17-27 121-4. See also Guo Zhenmin, *Shengsi yuye shihua*, 291–92; and Lu Yanghao, "Su sheng waihai ying fou zhengshou yuye linshi yingyeshui zhi shangque."

3. "Chongming xian zhengfu xunling," February 1935; "Chongming xian zhengfu, Jiangsu sheng Caizhengting zhishu Chongming yingyeshui zhengshouchu bugao," March 1935, IMH 17-27 121-4.

4. Mann, *Local Merchants and the Chinese Bureaucracy*, 248n237. The Chongming Business Tax Collection Office announced the provincial government's regulations for the collection of business taxes by brokers at the end of 1934. See "Jiangsu sheng Caizhengting zhishu Chongming yingyeshui zhengshouchu bugao," December 1934, IMH 17-27 121-4.

5. "Chongming xian yingyeshui zhengshouchu cheng Caizhengting wen," January 1935, IMH 17-27 121-4.

6. *ZSB*, March 25, 1935, 3; April 30, 1935, 3.

7. Leaders from the Renhe, Jinghe, and Hengshun lodges were in attendance (*DHZB*, May 5, 1935, 1; October 22, 1935, 1).

8. *DHZB*, August 21, 1935, 1.

9. *DHZB*, August 25, 1935, 2; November 7, 1935, 2.

10. *DHZB*, August 31, 1935, 2.

11. *DHZB*, June 6, 1936, 3.

12. Guo Zhenmin, *Shengsi yuye shihua*, 292–93; Lu Yanghao, "Su sheng waihai ying fou zhengshou yuye linshi yingye shui zhi shangque."

13. Ideally, the ward headman was to function as a submagistrate appointed by the province; see Duara, *Culture, Power, and the State*, 61.

14. For background information on the history of ward administration in Republican China, see ibid., 61–63, 83–84; and also Kuhn, "The Development of Local Government," 350–51.

15. *CMB*, April 2, 1934, 2; May 2, 1934, 2.

16. Wang Zenglu, "Jiangsu waihai caiwu xingzheng diaocha," 1.

17. *CMB*, January 1, 1933, 17.

18. Ibid.

19. Lu Yanghao, "Cong yuye guandian lun Shengsi de fen'ge," 3. See also *CMB*, January 17, 1934, 1; May 29, 1934, 2.

20. "Huyu banshichu cheng," March 24, 1934, IMH 17-26 134-4; "Haizhou yuye jishu chuanxisuo chunji shiye baogaoshu dier ce," August 1920, NHA 1038-2071; Zhongguo dier lishi dang'anguan, *Zhonghua minguo shi dang'an ziliao huibian*, vol. 5, 1, 7, 652–61. For a history of the coast guard in Republican China, see Han Yanlong, *Zhongguo jindai jingcha shi*, vol. 2, 479–90, 619–23.

21. "Yin xian Dongxiang Yongan gongsuo leshi bei," in Zhejiang sheng shuichanzhi bianzuan weiyuanhui, *Zhejiang sheng shuichanzhi*, 1147.

22. "Huyu banshichu zhuren Yuan Lianghua cheng," March 24, 1934, IMH 17-27 135-4. For another reference to coast guard collaboration with lodge organizations in fishing protection, see "Shiyebu Huyu banshichu zhuren Yuan Lianghua cheng," October 17, 1934, IMH 17-27 35-4.

23. "Zhejiang sheng weijing yi fa chengzhun beian ge tuanti choushou lougui yilanbiao," October 1934, IMH 17-27 35-4.

24. Wang Zenglu, "Jiangsu waihai caiwu xingzheng diaocha," 6–7.

25. Ibid., 9.

26. On the *tanpai* levy, see Duara, *Culture, Power, and the State*, 78–79.

27. Wang Zenglu, "Jiangsu waihai caiwu xingzheng diaocha," 2–7. Even though the Nationalist government made efforts to centralize these defense forces, local elites retained control of militias in areas throughout Jiangsu well into the 1930s; see Geisert, "Power and Society," 232–42.

28. Wang Zenglu, "Jiangsu waihai caiwu xingzheng diaocha," 7–8.

29. Guo Zhenmin, *Shengsi yuye shihua*, 293–94.

30. "Shiyebu zi," April 17, 1934, IMH 17-27 134-4; "Zhejiang sheng zhengfu zi," June 27, 1934, IMH 17-27 134-4. Zhao Cisheng was a native of Bizhu township's Sanshi village in Fenghua county (Fenghua shizhi bianzuan weiyuanhui, *Fenghua shizhi*, 966).

31. Kuhn, "The Development of Local Government," 351.

32. *DHZB*, December 31, 1935, 1.

33. "Yuye jingcha guicheng," June 27, 1931, IMH 17-27 2-3. See also Han Yanlong, *Zhongguo jindai jingcha shi*, vol. 2, 659–60.

34. *DHZB*, December 31, 1935, 1.

35. *DHZB*, April 17, 1935, 1.

36. *ZSB*, May 15, 1935, 5.

37. *DHZB*, May 17, 1935, 1.

38. "Shiyebu Huyu banshichu zhuren Yuan Lianghua cheng," June 24, 1935, IMH 17-27 139-1. See also *DHZB*, May 5, 1935, 1; and *ZSB*, May 4, 1935, 3. The Fishery Police did not register boats from Wenzhou and Taizhou, which paid fees to their own fishing organizations; see *DHZB*, December 31, 1935, 1.

39. *ZSB*, March 38, 1935, 3.

40. *DHZB*, May 11, 1934, 1–2.

41. It is possible that this shift was linked to strong summer monsoon winds, which push nutrient-rich, brackish waters from the Yangzi to the northeast; see Qiu Yongsong et al., "Runoff- and Monsoon-Driven

Variability of Fish Production in East China Seas," 24, 31; and Ge Quansheng, "1736 nian yilai Changjiang zhong xia you meiyu bianhua," 2797.

42. *DHZB*, May 17, 1935, 1.

43. *DHZB*, May 11, 1934, 1–2.

44. *DHZB*, April 9, 1936, 3.

45. *DHZB*, November 21, 1935, 1; "Shiyebu Huyu banshichu zhuren Yuan Lianghua cheng," June 24, 1935, IMH 17-27 139-1; *ZSB*, May 4, 1935, 3. On Shi Meiheng (also known as Shi Jichun), see "Zhejiang Yin xian yuhui zhiyuan lüli dan," 1934, IMH 17-27 127-2.

46. Jian, "Ningbo yujing zai Su jing Shengshan leizheng yumin qizhaofei jiufen ji," 10. See also Guo Zhenmin, *Shengsi yuye shihua*, 297–98.

47. "Jiangsu sheng ge xian chuanhu baojia buchong banfa," n.d.; "Diwu qu wailai chuanzhi biancha banfa," January 11, 1935, IMH 17-27 123-5. On the *baojia* collective security system in Jiangsu during the 1930s, see Geisert, "Power and Society," 185–88.

48. Cheng Tiyun, "Jiangsu waihai shan dao zhi," 17–18.

49. "Fujian Huian Fengwei bang yumin daibiao Liu Huifang deng cheng," July 29, 1935, IMH 17-27 35-5.

50. "Chongming xian yuhui Shengshan fenhui daidian," December 1935, IMH 17-27 128-2; "Jiangsu Chongming xian yuhui deng dian," January 4, 1935, IMH 17-27 138-3; *SCYK* 3, no. 3–4 (1935): 102–3; Jian, "Ningbo yujing zai Su jing Shengshan leizheng yumin qizhaofei jiufen ji," 10. See also *DHZB*, December 31, 1935, 1; and Guo Zhenmin, *Shengsi yuye shihua*, 297–98.

51. *DHZB*, May 5, 1935, 1; *ZSB*, May 6, 1935, 3; Ling Lingjiu and Cheng Tiyun, "Zhe Zhao zhuanyuan hua Sheng gui Zhe liyou zhi jiantao," 1.

52. These demands came from the All-Zhejiang Trade Association (Quan Zhe gonghui) and Shanghai's Ningbo, Dinghai, and Fenghua native-place associations (*SB*, June 6, 1935, 15; Lu Yanghao, "Pi Zhe sheng 'Ningbo,' 'Fenghua,' 'Dinghai' Hu tongxianghui qing hua Shengsi liedao gui Zhe shuo," 1; see also "Quan-Zhe gonghui daidian," June 25, 1935, IMH 17-27 28-3; and "Ningbo lü Hu tongxianghui deng dian," December 24, 1935, IMH 17-27 28-3).

53. *SB*, June 6, 1935, 15.

54. *ZSB*, May 6, 1935, 3; Ling Lingjiu and Cheng Tiyun, "Zhe Zhao zhuanyuan hua Sheng gui Zhe liyou zhi jiantao," 1.

55. *SB*, June 6, 1935, 15; Lu Yanghao, "Pi Zhe sheng 'Ningbo,' 'Fenghua,' 'Dinghai' Hu tongxianghui qing hua Shengsi liedao gui Zhe shuo," 1.

56. *DHZB*, February 15, 1936, 2.

57. *SB*, June 6, 1935, 15. Lu Yanghao, "Pi Zhe sheng 'Ningbo,' 'Feng-hua,' 'Dinghai' Hu tongxianghui qing hua Shengsi liedao gui Zhe shuo," 7.

58. "Han Guojun deng dian," January 31, 1936; "Du Yong deng dian," March 26, 1936, IMH 17-27 138-3.

59. Ling Lingjiu and Cheng Tiyun, "Zhe Zhao zhuanyuan hua Sheng gui Zhe liyou zhi jiantao," 2.

60. "Jiangsu shengli shuichan xuexiao xiaozhang Zhang Yulu cheng," July 1935, copy attached to "Xingzhengyuan mishuchu han," August 9, 1935, IMH 17-27 38-3.

61. Lu Yanghao, "Cong yuye guandian lun Shengsi de fen'ge," 23; "Jiangsu shengli shuichan xuexiao xiaozhang Zhang Yulu cheng," July 1935, IMH 17-27 28-3. Zhang Yulu graduated from the Jiangsu Provincial Fishery School before studying in Japan and had previously taught at the Zhejiang Provincial Fishery School ("Huiyuan lu," *Zhonghua minguo shuichan xuehui huibao* [1934], 108).

62. For an especially adamant statement of this claim, see "Chongming xian yuhui Shengshan fenhui changwu lishi Yang Youcai deng daidian," July 30, 1935, IMH 17-27 28-3.

63. "Jiangsu Chongming xian xian nonghui ganshizhang Huang Xian-lin deng cheng," July 24, 1935, IMH 17-27 28-3.

64. "Neizheng bu han," July 1935, copy attached to "Xingzhengyuan mishuchu han," August 30, 1935, IMH 17-27 38-3.

65. Chen Guofu's proposed solution was based on Section 8 of the central government's Provincial, Municipal, and County Boundary Survey Regulations (*Sheng, shi, xian kanjie tiaoli*), which spelled out procedures for resolving such disagreements ("Neizheng bu han," August 21, 1935; "Jiangsu sheng zhengfu han," n.d., both attached to "Neizhengbu mishu-chu han," August 31, 1935, IMH 17-27 38-3).

66. *DHZB*, February 15, 1936, 2. Chen Tongbai (1900–1984) studied at Saint John's University in his native Shanghai and Qinghua University in Tianjin before receiving an official scholarship to obtain his master's degree at the University of Washington in 1925. After coming back to China, Chen taught at the Jiangsu Provincial Fishery School and held positions in several fishery administration offices in Guangdong before taking over as head of the Zhejiang Provincial Fishery Experiment Station in 1935; see "Huiyuan lu," in *Zhonghua minguo shuichan xuehui huibao* (1934), 106; and Zhejiang sheng shuichanzhi bianzuan weiyuanhui, *Zhejiang sheng shuichanzhi*, 1017–18.

67. Lu Yanghao was a steering committee member in the Chongming County Nationalist Party branch's first plenary session in 1924 and head of its propaganda department; see "Chongming dangbu gongzuo diaocha shixiang yilan, Chongming xian dangwu zhengli weiyuanhui tianbao," August 1930, PHC 435-173; and also *CMB*, January 1, 1933, 6.

68. "Jiangsu sheng zhengfu yuan cheng" and "Zhejiang sheng zhengfu yuan cheng," both attached to "Xingzhengyuan mishuchu han," August 25, 1936, IMH 17-27 28-3; *SCYK* 3, no. 7 (1936): 79–80. See also Jin Gou, "Huanglong dao lishi yan'ge," 92.

69. "Xingzhengyuan mishuchu han," August 31, 1936; "Xingzhengyuan zhaoji Shengsi liedao huajie wenti an shenchahui," September 4, 1936, IMH 17-27 28-3. See also *SCYK* 3, no. 10 (1936): 64–66.

70. "Xingzhengyuan xunling," October 13, 1936, IMH 17-27 28-3; *SCYK* 3, no. 11 (1936): 77–78.

71. *SCYK* 3, no. 11 (1936): 79; 3, no. 12 (1936): 65–66; "Shengsi liedao shichatuan fan Hu," December 10, 1936, reproduced in Zhoushan shi dang'anguan, *"Shenbao" Zhoushan shiliao huibian*, 306–8. For a copy of the final report of Jiangsu's observatory mission to the Shengsi Islands, see "Shengsi liedao shidingtuan baogaoshu," December 31, 1936, SMA Q 464-568.

72. The committee was composed of the heads of the province's Civil Administration, Finance, Education, and Reconstruction departments. Special administrative supervisors from the Wenzhou region, Linhai county, and Ningbo, including Zhao Cisheng, were also members of the committee (*DHZB*, July 6, 1936, 2).

73. *DHZB*, July 14, 1936, 1.

74. Enterprises that failed to register could not request the Management Committee's protection or purchase salt used to preserve fish at reduced tax rates (*DHZB*, May 17, 1937, 2).

75. Shanghai yuyezhi bianzuan weiyuanhui, *Shanghai yuyezhi*, 549.

76. Hou Chaohai, "Woguo yuye gaikuang yu yuzheng sheshi fang-an," 42.

77. Ibid., 43.

78. Ibid., 43–44. See also Huang Zhenshi, "Jiu Shanghai de yushi," 231.

79. H.M., "Chinese Government Taking Control of Fish Industry," 187.

80. Huang Zhenshi, "Jiu Shanghai de yushi," 233.

81. Ibid., 232–33; Martin, *The Shanghai Green Gang*, 210–11; *Shanghai shi nianjian*, Q 71.

82. K. C. Lin, "Shanghai's New Fish Market," 607–8.

83. H.M., "Chinese Government Taking Control of Fish Industry," 187.

84. *Shanghai shi nianjian*, Q 71–77.

85. For the Nationalist government's concern about revenues lost as a result of Japanese boats that avoided payment of customs duties in Shanghai, see "Caizhengbu zi," February 14, 1933, IMH 17-27 122-5.

86. *SCYK* 3, no. 5–6 (1936): 104.

87. Li Shihao and Qu Ruoqian, *Zhongguo yuye shi*, 98–100. See also IMH 17-27 38-4.

88. "Huazhong zhenxing gongsi ji qi guanxi gongsi zhi yanjiu," 1944, in Zhongguo dier lishi dang'anguan, *Zhonghua minguo shi dang'an ziliao huibian*, vol. 5, 2, *fu lu*, 1094. See also Huang Meizhen, *Ri wei dui Huazhong lunxian qu jingji de lüeduo yu tongzhi*, 396. On the military situation in Shanghai and the Lower Yangzi during 1937, see Williamsen, "The Military Dimension," 142–44.

89. Guo Zhenmin, *Shengsi yuye shihua*, 78, 272; Zhoushan shi dang'anguan, *"Shenbao" Zhoushan shiliao huibian*, 331; Zhoushan yuzhi bianxie zu, *Zhoushan yuzhi*, 32.

90. Okamoto Shoichi, *Man-Shi no suisan jijō*, 610–11.

91. For the Central China Marine Products Company, see Coble, *Chinese Capitalists in Japan's New Order*, 58, 79. See also Okamoto Nobuo, *Kindai gyogyō hattatsu shi*, 417; and Okamoto Shoichi, *Man-Shi no suisan jijō*, 608–12, 910–13.

92. *DHMB*, August 23, 1938, 2; August 27, 1938, 2.

93. "Huazhong zhenxing gongsi ji qi guanxi gongsi zhi yanjiu," 1944, in Zhongguo dier lishi dang'anguan, *Zhonghua minguo shi dang'an ziliao huibian*, vol. 5, 2, *fu lu*, 1094.

94. Huang Meizhen, *Ri wei dui Huazhong lunxian qu jingji de lüeduo yu tongzhi*, 398. According to one Japanese source, almost 80 percent of the catch marketed in Shanghai during 1939 was brought to port by Chinese boats, and only 20 percent was landed by vessels directly operated by the Central China Marine Products Company (Okamoto Shoichi, *Man-Shi no suisan jijō*, 913).

95. *DHMB*, July 20, 1938, 2.

96. *DHMB*, August 23, 1938, 2; September 15, 1938, 2.

97. *DHMB*, August 23, 1938, 2; September 15, 1938, 2.

98. Zhejiang sheng ge xian jianshe gongzuo taolunhui mishuchu, *Zhejiang sheng ge xian jianshe gongzuo taolunhui huiyi baogao*, vol. 2, 381.

99. "Qian Ningbo yuye guanli ju zongwu He Yu cheng," May 12, 1942, IMH 28-03-05 8-2. See also Okamoto Shoichi, *Man-Shi no suisan jijō*, 610.

100. Zhejiang sheng ge xian jianshe gongzuo taolunhui mishuchu, *Zhejiang sheng ge xian jianshe gongzuo taolunhui huiyi baogao*, vol. 2, 404, 419.

101. *DHMB*, July 29, 1938, 2.

102. Zhejiang sheng ge xian jianshe gongzuo taolunhui mishuchu, *Zhejiang sheng ge xian jianshe gongzuo taolunhui huiyi baogao*, vol. 2, 373, 375, 386, 402, 404, 432–33, 444, 456–67.

103. "Ningbo, Dinghai yushi yuchuan yumin diaocha baogaoshu," August 31, 1938, ZPA 33-2-314.

104. *DHMB*, August 11, 1938, 2.

105. Zhejiang sheng ge xian jianshe gongzuo taolunhui mishuchu, *Zhejiang sheng ge xian jianshe gongzuo taolunhui huiyi baogao*, 36–41; "Jiuji ben sheng yuye an," April 1940, AH 064-1443; Hou Chaohai et al., "Qing dingding kangzhan shiqi tuijin quanguo yuye hezuo shiye an," May 1941, AH 124-325.

106. Li Xingjie, "Zhejiang sheng yuzheng zhi huigu yu qianzhan," 8–9. For an overview of military developments in Zhejiang during the war, see Zhang Genfu, *Kangzhan shiqi Zhejiang sheng renkou qianyi yu shehui yingxiang*, 26–29.

107. Li Xingjie, "Zhejiang sheng yuzheng zhi huigu yu qianzhan," 5.

108. Zhoushan yuzhi bianxie zu, *Zhoushan yuzhi*, 53.

109. "Huazhong shuichan gufen youxian gongsi di shiyi qi dingqi gudong dahuiyi an," March 1945, IMH 28-03-05 007-2.

110. Tsutsui, "Landscapes in the Dark Valley," 302–3.

Conclusion

1. Shapiro, *Mao's War Against Nature*, 4.

2. Zhao Yizhong, "Zhonghua renmin gongheguo jianguo chuqi huifu Zhoushan yuye shengchan de ji ge licheng," 81–82; Zhoushan yuzhi bianxie zu, *Zhoushan yuzhi*, 332–33.

3. Guo Zhenmin, *Shengsi yuye shihua*, 349–52; Zhao Yizhong, "Zhonghua renmin gongheguo jianguo chuqi huifu Zhoushan yuye shengchan de ji ge licheng," 89–90; Zhoushan yuzhi bianxie zu, *Zhoushan yuzhi*, 299–302.

4. Guo Zhenmin, *Shengsi yuye shihua*, 183–87; Zhao Yizhong, "Zhonghua renmin gongheguo jianguo chuqi huifu Zhoushan yuye shengchan de ji ge licheng," 84; Zhoushan yuzhi bianxie zu, *Zhoushan yuzhi*, 254–56.

5. Zhao Yizhong, "Zhonghua renmin gongheguo jianguo chuqi huifu Zhoushan yuye shengchan de ji ge licheng," 91–93; Zhoushan yuzhi bianxie zu, *Zhoushan yuzhi*, 115–17. For Republican-period proposals, see "Quanguo shuichan shuxue jiguan gaikuang," 90, in *Zhonghua minguo shuichan xuehui huibao* (1934), SMA Y 4-1-225.

6. Solecki, *Economic Aspects of the Fishing Industry in Mainland China*, 78–79.

7. Huang Junming, "Daishan yuye lishi tedian tantao," 50–51; Zhao Yizhong, "Zhonghua renmin gongheguo jianguo chuqi huifu Zhoushan yuye shengchan de ji ge licheng," 93.

8. Solecki, *Economic Aspects of the Fishing Industry in Mainland China*, 106–7.

9. Guo Zhenmin, *Shengsi yuye shihua*, 223–37; Liu Zhengquan, "Shengsi liedao jubian," 129–30.

10. For biographical information on these fishery experts' official responsibilities during the post–1949 period, see Shanghai yuyezhi bianzuan weiyuanhui, *Shanghai yuyezhi*, 547–49; and Zhejiang sheng shuichanzhi bianzuan weiyuanhui, *Zhejiang sheng shuichanzhi*, 1008–9, 1015–16.

11. Guo Zhenmin, *Shengsi yuye shihua*, 284–86; Zhao Yizhong, "Zhonghua renmin gongheguo jianguo chuqi huifu Zhoushan yuye shengchan de ji ge licheng," 85–86, 90.

12. Zhao Yizhong, "Zhonghua renmin gongheguo jianguo chuqi huifu Zhoushan yuye shengchan de ji ge licheng," 86–87.

13. Ibid., 85.

14. Zhang Lixiu and Bi Dingbang, *Zhejiang dangdai yuye shi*, 27.

15. Qiu Yongsong et al., "Runoff- and Monsoon-Driven Variability of Fish Production in East China Seas."

16. Ibid., 85.

17. Ibid., 82.

18. Guo Zhenmin, *Shengsi yuye shihua*, 302–3; Zhao Yizhong, "Zhonghua renmin gongheguo jianguo chuqi huifu Zhoushan yuye shengchan de ji ge licheng," 84, 96.

19. Guo Zhenmin, *Shengsi yuye shihua*, 285.

20. "Yi jiu liu san nian quanguo shuichan gongzuo huiyi jiyao," December 1963, SMA B255-1-333.

21. "Zhejiang sheng renmin zhengfu Nonglinting shuichanju han," March 26, 1954; "Taizhou zhuanqu yumin xiehui weiyuanhui," April 10, 1954, ZPA 007-6-20.

22. For sources related to Japanese boats' continued fishing activities in waters off the Zhoushan Archipelago during the early PRC period, see "Guoying Shanghai shuichan gongsi yulun hanwei haifang he baohu yuchang qingkuang," December 1950–October 1952, SMA B255-1-37; and "Donghai yuqu dui Ri yulun douzheng de qingkuang he chengli Donghai yuye shengchan lianhe zhihuibu yijian," March 1959, SMA B-1-222. On the revival of the Japanese fishing industry during the late 1940s and early 1950s, see Scheiber, *Inter-Allied Conflicts and Ocean Law*.

23. Song, "China's Ocean Policy," 989–90.

24. "Donghai qu yuye shengchan lianhe zhihuibu guanyu yunlun hui ling jinru jinyuqu shengchan de tongbao," March 31, 1961, ZPA J122-16-13.

25. Solecki, *Economic Aspects of the Fishing Industry in Mainland China*, 118.

26. "Guanyu Lüsi yuchang de yixie cailiao," March 1, 1968, 5, in Zhejiang sheng, Shuichanting, *Lüsi yuchang shengchan ziliao*.

27. Solecki, *Economic Aspects of the Fishing Industry in Mainland China*, 118–19, 141.

28. Qiu Yongsong et al., "Runoff- and Monsoon-Driven Variability of Fish Production in East China Seas."

29. Guo Zhenmin, *Shengsi yuye shihua*, 126–27, 199–214; Huang Junming and Zhang Mingquan, "Daishan yuye jifanhua shihua," 120; Shang Hong, "Haiyang yuye ziyuan baohu yu guanli wenti de tantao"; Zhoushan yuzhi bianxie zu, *Zhoushan yuzhi*, 52–67; Zhang Lixiu and Bi Dingbang, *Zhejiang dangdai yuye shi*, 86–91.

30. Libecap, *Contracting for Property Rights*, 27.

31. Smith, *Scaling Fisheries*, 336.

32. Weller, *Discovering Nature*, 107.

33. The formation of proprietary claims to Zhoushan's fishing grounds corresponds with economists' prediction that degree of "definition and enforcement activity" will increase with the value of assets and the probability of losing the right to use them (Anderson and Hill, "From Free Grass to Fences").

34. McEvoy, *The Fisherman's Problem*, 99.

35. As Theodore Bestor (*Tsukiji*, 15–16) observes in his ethnography of a Tokyo fish market, without institutional structure "systematic economic activity—that is, a market—could not exist."

36. Lamley, "Lineage Feuding in Southern Fujian and Eastern Guangdong Under Qing Rule," 57.

37. Ibid., 47, 57–58.

38. Leong, *Migration and Ethnicity*, 13, 39, 56.
39. Perry, *Rebels and Revolutionaries in North China*, 47.
40. See Duara, *Culture, Power, and the State*.
41. Elvin, "Three Thousand Years of Unsustainable Growth," 11, 21–22, 46.
42. McNeill, *Something New Under the Sun*, 361.
43. Shang Hong, "Haiyang yuye ziyuan baohu yu guanli wenti de tantao," 825–26; Zheng Pingsheng, "Haiyang ziyuan guanli cuoshi chutan," 818.

Works Cited

Acheson, James M. "Anthropology of Fishing." *Annual Review of Anthropology* 10 (1981).

———. *Capturing the Commons: Devising Institutions to Manage the Maine Lobster Industry*. Hanover: University Press of New England, 2003.

Anderson, Terry L., and P. J. Hill. "From Free Grass to Fences: Transforming the Commons in the American West." In *Managing the Commons*, ed. John A. Baden and Douglas S. Noonan. Bloomington: Indiana University Press, 1998.

Antony, Robert J. *Like Froth Floating on the Sea: The World of Pirates and Seafarers in Late Imperial South China*. Berkeley: Institute of East Asian Studies, University of California, Berkeley, 2003.

Baden, John A. "A New Primer for the Management of Common-Pool Resources and Public Goods." In *Managing the Commons*, ed. idem and Douglas S. Noonan. Bloomington: University of Indiana Press, 1998.

Baden, John A., and Douglas S. Noonan. "Preface: Overcoming the Tragedy." In *Managing the Commons*, ed. idem and Douglas S. Noonan. Bloomington: Indiana University Press, 1998.

Bankoff, Greg. "Coming to Terms with Nature: State and Environment in Maritime Southeast Asia." *Environmental History Review* 19 (1995).

Bestor, Theodore C. *Tsukiji: The Fish Market at the Center of the World*. Berkeley: University of California Press, 2004.

Bolster, W. Jeffrey. "Opportunities in Marine Environmental History." *Environmental History* 11, no. 3 (2006).

———. "Putting the Ocean in Atlantic History: Maritime Communities and Marine Ecology in the Northwest Atlantic, 1500–1800." *American Historical Review* 113, no. 1 (2008).

Boorman, Howard L. *Biographical Dictionary of Republican China*. 5 vols. New York: Columbia University Press, 1967–79.

Buoye, Thomas M. *Manslaughter, Markets, and Moral Economy: Violent Disputes over Property Rights in Eighteenth-Century China*. Cambridge: Cambridge University Press, 2000.

Cadigan, Sean T., and Jeffrey A. Hutchings. "Nineteenth-Century Expansion of the Newfoundland Fishery for Atlantic Cod: An Exploration of Underlying Causes." In *The Exploited Seas: New Directions for Marine Environmental History*, ed. Tim D. Smith, Poul Holm, and David J. Starkey. St. John's: International Maritime Economic History Association/Census of Marine Life, 2001.

Cai Yong. "Zhoushan zui zao de yi jia yiyuan: gongli Cunji yiyuan" (Zhoushan's earliest hospital: the public Cunji hospital). *Zhoushan wenshi ziliao* 1 (1991).

Cao Erhui. "Su sheng waihai jiushi yuye zhi" (Record of old-style offshore fisheries in Jiangsu province). *Jiangsu yanjiu* 1, no. 3 (1935).

Chang, William Y. B., and George King. "Centennial Climate Changes in the Yangtze River Delta." *Climate Research* 4 (1994).

Chen Guoqiang and Cai Yongzhe. *Chongwu renleixue diaocha* (Anthropological investigation of Chongwu). Huian: Fujian jiaoyu chubanshe, 1990.

Chen Musen et al. "Daishan xingshi tan" (A discussion of Daishan's surname groups). *Daishan wenshi ziliao* 4 (1992).

Chen Shanqing, ed. *Dongchen cunzhi waibian* (Dongchen village gazetteer, additional sections). Xiangshan xian, 1999.

Chen, Ya-Qu, and Shen Xin-Qiang. "Changes in the Biomass of the East China Sea Ecosystem." In *The Large Marine Ecosystems of the Pacific Rim: A Report of a Symposium Held in Qingdao, People's Republic of China, 8–11 October 1994*, ed. Q. Tang and K. Sherman. Gland, Switzerland: A Marine Conservation and Development Report, IUCN, 1995.

Chen Ya-Qu and Xin-Qiang Shen. "Changes in the Biomass of the East China Sea Ecosystem." In *Large Marine Ecosystems of the Pacific Rim: Assessment, Sustainability, and Management*, ed. Kenneth Sherman and Qisheng Tang. Malden, MA: Blackwell Science, 1999.

Chen, Yixin. "The Guomindang's Approach to Rural Socioeconomic Problems: China's Rural Cooperative Movement, 1918–1949." PhD diss., Washington University, Saint Louis, 1995.

Chen Yuxin. "Zhoushan bingchang" (Zhoushan's icehouses). In *Wenshi tiandi*, ed. Zhoushan shi zhengxie wenshi he xuexi wei. Beijing: Wenjin chubanshe, 2003.

Cheng, Linsun. *Banking in Modern China: Entrepreneurs, Professional Managers, and the Development of Chinese Banks, 1897–1937.* Cambridge: Cambridge University Press, 2003.

Cheng Tiyun. "Jiangsu waihai shan dao zhi" (Record of Jiangsu's offshore islands). *Jiangsu yanjiu* 1, no. 6 (1935).

Churchhill, R. R., and A. V. Lowe. *The Law of the Sea.* Manchester: Manchester University Press, 1999.

Coble, Parks M. *Chinese Capitalists in Japan's New Order: The Occupied Lower Yangzi, 1937–1945.* Berkeley: University of California Press, 2003.

———. *Facing Japan: Chinese Politics and Japanese Imperialism.* Cambridge: Harvard University East Asian Monographs, 1991.

Cronon, William. *Nature's Metropolis: Chicago and the Great West.* New York: Norton, 1991.

Cushing, D. H. *The Provident Sea.* Cambridge: Cambridge University Press, 1988.

Daishan zhenzhi (Daishan market-town gazetteer). 1921.

de Groot, J. J. M. *The Religious System of China.* 6 vols. Leiden: Brill, 1892–1910.

Dhokalia, R. H. *The Codification of Public International Law.* Manchester: Manchester University Press, 1970.

Ding Fanglong and Guan Baoren. "Jiefang qian Daishan shangye gaikuang" (The general situation of commerce in Daishan before 1949). *Daishan wenshi ziliao* 4 (1992).

———. "'Penglai shi jing' tan" (A discussion of *Eight Views of Penglai*). *Daishan wenshi ziliao* 4 (1992).

Dinghai tingzhi (Dinghai sub-prefecture gazetteer). 1885.

Dinghai xianzhi (Dinghai county gazetteer). 1924.

Dinghai xianzhi bianzuan weiyuanhui. *Dinghai xianzhi* (Dinghai county gazetteer). Hangzhou: Zhejiang renmin chubanshe, 1994.

Du Haogeng. "Moyu bulong wenti de jiantao" (A brief discussion of the problem of cuttlefish cages). *Shuichan yuekan* 3, no. 8 (1936).

Duara, Prasenjit. *Culture, Power, and the State: Rural North China, 1900–1942*. Stanford: Stanford University Press, 1988.

Economy, Elizabeth C. *The River Runs Black: The Environmental Challenge to China's Future*. Ithaca: Cornell University Press, 2004.

Elvin, Mark. *The Retreat of the Elephants: An Environmental History of China*. New Haven: Yale University Press, 2004.

———. "Three Thousand Years of Unsustainable Growth: China's Environment from Archaic Times to the Present." *East Asian History* 6 (1993).

Fang Changsheng, ed. *Zhejiang sheng minjian wenxue jicheng: Zhoushan shi geyao yanyu juan* (Zhejiang province popular literature collection: Zhoushan municipality folk songs and sayings section). Beijing: Zhongguo minjian wenyi chubanshe, 1989.

———. *Zhoushan minsu wenxue yanjiu* (Research on Zhoushan popular literature). Beijing: Zhongguo wenshi chubanshe, 2005.

Fang Changsheng and Wang Daoxing. *Zhoushan fengsu* (Zhoushan customs). Dinghai: Zhoushan shi wenyijie lianhehui, 1988.

Feeny, David; Fikret Berkes; Bonnie J. McCay; and James M. Acheson. "The Tragedy of the Commons: Twenty-Two Years Later." *Human Ecology* 18 (1990).

Fenghua shizhi bianzuan weiyuanhui. *Fenghua shizhi* (Fenghua municipal gazetteer). Beijing: Zhonghua shuju, 1994.

"The Fishing Industry in Kiangsu." *Chinese Economic Journal* 3 (1928).

Ford, John D. *An American Cruiser in the Far East*. New York: A. S. Barnes, 1898.

Fortune, Robert. *Three Years' Wanderings in the Northern Provinces of China*. London: John Murray, 1847.

Frawley, Kevin. "Evolving Visions: Environmental Management and Nature Conservation in Australia." In *Australian Environmental History: Essays and Cases*, ed. Stephen Dovers. Oxford: Oxford University Press, 1994.

Friedman, Sara L. *Intimate Politics: Marriage, the Market, and State Power in Southeastern China*. Cambridge: Harvard University Asia Center, 2006.

Fu Guozhang. "Yuhang yu bingxian" (Fish brokers and ice boats). *Shengsi wenshi ziliao* 2 (1989).

Gan Yuli. "Jiang Zhe waihai yuye xiankuang" (The current state of offshore fisheries in Jiangsu and Zhejiang). *Fuxing yuekan* 1, no. 3 (1932).

Ge Quansheng. "1736 nian yilai Changjiang zhong xia you meiyu bianhua" (Changes in the *meiyu* rains in the middle and lower Yangzi since 1736). *Kexue tongbao* 52, no. 23 (2007).

Geisert, Bradley K. "Power and Society: The Kuomintang and Local Elites in Kiangsu Province, China, 1924–1937." PhD diss., University of Virginia, 1979.

Gerth, Karl. *China Made: Consumer Culture and the Creation of the Nation.* Cambridge: Harvard University Asia Center, 2003.

Geyao. "Zhoushan nanzi de chuantong zhiye" (Traditional occupations of males in Zhoushan). In *Wenshi tiandi*, ed. Zhoushan shi zhengxie wenshi he xuexi wei. Beijing: Wenjin chubanshe, 2003.

Golas, Peter J. "Early Ch'ing Guilds." In *The City in Late Imperial China*, ed. G. William Skinner. Stanford: Stanford University Press, 1977.

Gong Yu and Huang Zhiguo. "Zhoushan de 'Mazu miao': Dongshan Yangfu gong" (Zhoushan's Mazu Temple: Dongsha's Yang Fu Temple). In *Wenshi tiandi*, ed. Zhoushan shi zhengxie wenshi he xuexi wei. Beijing: Wenjin chubanshe, 2003.

Goodman, Bryna. "The Native Place and the City: Immigrant Consciousness and Organization in Shanghai, 1853–1927." PhD diss., Stanford University, 1990.

———. *Native Place, City, and Nation: Regional Networks and Identities in Shanghai, 1853–1937.* Berkeley: University of California Press, 1995.

Gordon, H. Scott. "The Economic Theory of a Common Property Resource: The Fishery." *Journal of Political Economy* 62 (1954).

Gu Mingsheng, trans. *Shuichanxue xinbian* (Fishery studies, new edition). Shanghai: Shanghai kexue shuju, 1911.

Gu Zongjian. "Weijingtan yinggao" (Weijingtan hard biscuits). *Daishan wenshi ziliao* 1 (1986).

Guan Pengwan. *Shuichanxue da yi* (The essentials of fishery studies). Shanghai: Shangwu yinshuguan, 1919.

Guo Jizhong. "Kanmen san qianzhuang" (Three native banks in Kanmen). *Yuhuan wenshi ziliao* 4 (1988).

Guo Qiyun et al. "1873–2000 nian Dongya xiajifeng bianhua de yanjiu" (Studies on variation in the East Asian summer monsoon, 1873–2000). *Daqi kexue* 28, no. 2 (2004).

Guo Zhenmin. *Shengsi yuye shihua* (Historical discussion of Shengsi's fisheries). Beijing: Haiyang chubanshe, 1995.

256   Works Cited

Han Yanlong. *Zhongguo jindai jingcha shi* (History of police in modern China). 2 vols. Beijing: Shehui kexue wenxian chubanshe, 1999.

Hardin, Garrett. "The Tragedy of the Commons." *Science* 162 (1968).

He Yunyu and Wei Xiyan. "Jianguo qian Chongwu yuye gaikuang he yu yahang sheng'ai" (The general situation of Chongwu's fisheries and the rise and decline of fish brokerages before 1949). *Huian wenshi ziliao* 6 (1989).

Himeda Mitsuyoshi. "Chūgoku kindai gyogyō shi no hitokoma: Kampō hachinen Kin ken no gyomin tōsō o megutte" (A page in the history of China's modern fishing industry: Yin county fishermen's struggle in the eighth year of the Xianfeng reign). In *Kindai Chūgoku nōson shakai shi kenkyū*, ed. Tōkyō kyōiku daigaku, Tōyō shigaku kenkyūshitsu, Ajia shi kenkyūkai, Chūgoku kindai shi kenkyūkai. Tokyo: Daian, 1967.

H.M. "Chinese Government Taking Control of Fish Industry." *Far Eastern Survey* 5, no. 17 (1936).

Ho, Ping-ti. *Studies on the Population of China, 1368–1953.* Cambridge: Harvard University Press, 1959.

Hoffmann, Richard C. "Economic Development and Aquatic Ecosystems in Medieval Europe." *American Historical Review* 101, no. 3 (1996).

Hou Chaohai. "Fakan ci" (Introduction to the first issue). *Zhonghua minguo shuichan xuehui huibao* (1934).

———. "Shiyebu Jiang-Zhe qu yuye gaijin weiyuanhui chengli shi yuzheng zhi sheshi tan" (A discussion of the history of the establishment of the Jiangsu-Zhejiang Region Fishery Reform Committee and the organization of fishery administration). *Zhonghua minguo shuichan xuehui huibao* (1934).

———. "Woguo yuye gaikuang yu yuzheng sheshi fang'an" (The general condition of our country's fisheries and plans for the organization of fishery administration). *Shuichan yuekan* 3, no. 1 (1934).

Howell, David L. *Capitalism from Within: Economy, Society, and the State in a Japanese Fishery.* Berkeley: University of California Press, 1995.

Hu Juntai. "Zhejiang shuichan zhi wojian" (My opinions on Zhejiang's fisheries). *Zhejiang shengli shuichanke zhiye xuexiao xiaokan* (1929).

Huang Junming. "Daishan yuye lishi tedian tantao" (An inquiry into the special characteristics of the history of Daishan's fisheries). *Daishan wenshi ziliao* 1 (1986).

Huang Junming and Zhang Mingquan. "Daishan yuhangzhan qianshuo" (An elementary introduction to Daishan's fish brokers). *Daishan wenshi ziliao* 4 (1992).

——. "Daishan yuye jifanhua shihua" (Historical discussion of the mechanization of Daishan's fisheries). *Daishan wenshi ziliao* 4 (1992).

Huang Meizhen. *Ri wei dui Huazhong lunxian qu jingji de lüeduo yu tongzhi* (Japan's economic exploitation and control of occupied areas in central China). Beijing: Shehui kexue wenxian chubanshe, 2005.

Huang, Philip C. C. "Between Informal Mediation and Formal Adjudication: The Third Realm of Qing Civil Justice." *Modern China* 19, no. 3 (1993).

Huang Zhenshi. "Jiu Shanghai de yushi" (Old Shanghai's fish market). In *Shanghai wenshi ziliao cungao huibian*, ed. Shanghai shi zhengxie wenshi ziliao weiyuanhui. Shanghai: Shanghai guji chubanshe, 1980.

Imperial Maritime Customs. *Decennial Reports on the Trade, Navigation, Industries, etc. of the Ports Open to Foreign Commerce in China, and on the Condition and Development of the Treaty Port Provinces, 1892–1901.* Shanghai: Inspectorate General of Customs, 1904, 1906.

——. *Reports on the Trade of the Treaty Ports of China: Taichow.* Shanghai: Inspectorate General of Customs, 1881.

——. *Returns of Trade and Trade Reports.* Shanghai: Inspectorate General of Customs, 1906.

Iriye, Akira. *After Imperialism: The Search for a New Order in the Far East, 1921–1931.* Cambridge: Harvard University Press, 1965.

——. *China and Japan in the Global Setting.* Cambridge: Harvard University Press, 1992.

Iversen, Edwin S. *Living Marine Resources: Their Utilization and Management.* New York: Chapman and Hall, 1996.

Jackson, Jeremy B. C., et al. "Historical Overfishing and the Recent Collapse of Coastal Ecosystems." *Science* 293 (2001).

Jennings, Simon; Michel J. Kaiser; and John D. Reynolds. *Marine Fisheries Ecology.* Malden, MA: Blackwell Science, 2001.

Jian (pseud.). "Ningbo yujing zai Su jing Shengshan leizheng yumin qizhaofei jiufen ji" (A record of the dispute over the collection of permit fees from fishermen by the Ningbo Fishery Police in Jiangsu's territory at Shengshan). *Jiangsu yanjiu* 1, no. 8 (1935).

Jiang Bin and Jin Tao. *Donghai daoyu wenhua yu minsu* (The culture and popular customs of islands in the East China Sea). Shanghai: Shanghai wenyi chubanshe, 2005.

"Jiang Zhe yanhai zhi moyu yuchang ji yumin shenghuo" (Cuttlefish fishing grounds and the life of fishermen in coastal Jiangsu and Zhejiang). *Gongshang banyuekan* 5, no. 5 (1933).

Jie Tianhai. "Caiyuan yuhang xingshuai ji" (A record of the rise and decline of Caiyuan's fish brokers). *Shengsi wenshi ziliao* 2 (1989).

Jin Gou. "Huanglong dao lishi yan'ge" (The historical development of Huanglong Island). *Shengsi wenshi ziliao* 2 (1991).

Jin Jifu. "Zhejiang Tai shu shuichan gaikuang" (The general condition of marine products in Taizhou, Zhejiang). *Zhejiang sheng jianshe yuekan* 7, no. 9 (1934).

Jin Li. "Hengjie yushi shihua" (Historical discussion of the Hengjie fish market). *Daishan wenshi ziliao* 1 (1986).

Jin Tao. "'Shengsi yumin fengsu' kao" (An investigation of Shengsi fishermen's customs). *Shengsi wenshi ziliao* 1 (1989).

Jin Xinheng. "Lun yupin xiaolu tuiguang fangfa" (A discussion of methods for expanding the marketing of fish products). *Zhejiang shengli shuichanke zhiye xuexiao xiaokan* (1929).

Jin Zhaohua. "Zhejiang shuichan jianshe wenti zhi jiantao" (A preliminary inquiry into the problem of the reconstruction of Zhejiang's fisheries). *Zhejiang sheng jianshe yuekan* 7, no. 9 (1934).

Jin Zhiquan. "Yu shichang zhi jianshe" (The reconstruction of fish markets). *Zhejiang shengli shuichanke zhiye xuexiao xiaokan* (1929).

———. "Zhejiang yuye zhi xianzai ji jianglai zhi qushi" (Present and future trends in Zhejiang's fisheries). *Zhongguo jianshe* 3, no. 4 (1931).

Jordan, Donald A. *Chinese Boycotts Versus Japanese Bombs: The Failure of China's "Revolutionary Diplomacy," 1931–1932*. Ann Arbor: University of Michigan Press, 1991.

Josephson, Paul R. *Resources Under Regimes: Technology, Environment, and the State*. Cambridge: Harvard University Press, 2004.

Kemp, Peter, ed. *The Oxford Companion to Ships and the Sea*. Oxford: Oxford University Press, 1976.

Kibesaki Osamu. "Fundamental Studies on Structure and Effective Management of the Demersal Fish Resources in the East China and the Yellow Sea." *The Investigations of Demersal Fish Resources in the East China and Yellow Seas* 5 (1960).

King, Frank H. H. "Pricing Policy in a Chinese Fishing Village." *Journal of Oriental Studies* 1, no. 1 (1954).

Kirby, William C. "China Unincorporated: Company Law and Business Enterprise in Twentieth-Century China." *Journal of Asian Studies* 54, no. 1 (1995).

———. "Engineering China: The Birth of the Developmental State, 1928–1937." In *Becoming Chinese: Passages to Modernity and Beyond*, ed. Wen-hsin Yeh. Berkeley: University of California Press, 2000.

Kobayashi Sōichi. *Shina no janku* (Chinese junks). Tokyo: Dai Nihon Tōkyō kaiyō shōnendan, 1942.

Köll, Elisabeth. *From Cotton Mill to Business Empire: The Emergence of Regional Enterprises in Modern China*. Cambridge: Harvard University Asia Center, 2003.

Koppes, Clayton R. "Efficiency, Equality, Esthetics: Shifting Themes in American Conservation." In *The Ends of the Earth: Perspectives on Modern Environmental History*, ed. Donald Worster. Cambridge: Cambridge University Press, 1988.

Kōsō shō Sekkō shō suisangyō chōsa hōkoku (Investigation report on Jiangsu province and Zhejiang province's marine products industry). Taihoku: Taiwan sōtokufu, Shokusanchō, Kōshoka, 1924.

Kuhn, Philip A. "The Development of Local Government." In *The Cambridge History of China*, vol. 13, pt. 2, 329–60. Cambridge: Cambridge University Press, 1986.

———. "Local Self-Government Under the Republic: Problems of Control, Autonomy, and Mobilization." In *Conflict and Control in Late Imperial China*, ed. Frederic Wakeman, Jr., and Carolyn Grant. Berkeley: University of California Press, 1975.

———. *Rebellion and Its Enemies in Late Imperial China: Militarization and Social Structure, 1769–1864*. Cambridge: Harvard University Press, 1970.

———. *Soulstealers: The Chinese Sorcery Scare of 1769*. Cambridge: Harvard University Press, 1990.

———. "Toward a Historical Ecology of Chinese Migration." In *The Chinese Overseas*, vol. 1, ed. Liu Hong. London: Routledge Library of Modern China, 2006.

Lamley, Harry J. "Lineage Feuding in Southern Fujian and Eastern Guangdong Under Qing Rule." In *Violence in China: Essays in Culture and Counterculture*, ed. Jonathan N. Lipman and Stevan Harrell. Albany: State University of New York Press, 1990.

Lavely, William; James Lee; and Wang Feng. "Chinese Demography: The State of the Field." *Journal of Asian Studies* 50, no. 1 (1990).

"The Law of Territorial Waters." *American Journal of International Law* 23, no. 2, Supplement: Codification of International Law (1929).

Leong, Sow-Theng. *Migration and Ethnicity in Chinese History: Hakkas, Pengmin, and Their Neighbors*. Ed. Tim Wright; Introduction and maps by G. William Skinner. Stanford: Stanford University Press, 1997.

Li Guoqi. *Zhongguo xiandaihua de quyu yanjiu: Min Zhe Tai diqu, 1860–1916* (Regional research on China's modernization: Fujian, Zhejiang, and Taiwan, 1860–1916). Taibei: Zhongyang yanjiuyuan, Jindaishi yanjiu-suo, 1982.

Li Jian. *Shanghai de Ningboren* (Shanghai's Ningbo people). Shanghai: Shanghai renmin chubanshe, 2000.

Li Rongsheng. *Zhongguo shuichan dili* (China's marine products geography). Beijing: Nongye chubanshe, 1985.

Li Shihao. *Zhongguo haiyang yuye xianzhuang ji qi jianshe* (The current state of China's marine fisheries and their reconstruction). Shanghai: Shangwu yinshuguan, 1936.

Li Shihao and Qu Ruoqian. *Zhongguo yuye shi* (History of China's fisheries). Shanghai: Shangwu yinshuguan, 1937.

Li Shiting. "'Sucheng' xiaokao" (A quiz on common sayings). In *Wenshi tiandi*, ed., Zhoushan shi zhengxie wenshi he xuexi wei. Beijing: Wenjin chubanshe, 2003.

Li Xingjie. "Zhejiang sheng yuzheng zhi huigu yu qianzhan" (A review of Zhejiang province's fishery policy and its future prospects). *Shuichan yuekan* 1, no. 5 (1946).

Li Yaohui. "Jiangsu zhi yuye gaikuang" (The general condition of Jiangsu's fisheries). *Zhongguo jianshe* 3, no. 4 (1931).

Li Yixiang. *Jindai Zhongguo yinhang yu qiye fazhan de guanxi* (The relationship between modern Chinese banks and enterprise development). Taibei: Dongda tushu gongsi, 1997.

Libecap, Gary D. *Contracting for Property Rights*. Cambridge: Cambridge University Press, 1989.

Lin, K. C. "Shanghai's New Fish Market." *Chinese Economic Journal* 15 (1934).

Lin Maochun and Wu Yuqi. *Yin xian yuye diaocha baogao* (Investigation report on Yin county's fisheries). Zhejiang shuichan shiyanchang huikan 2, no. 2. Dinghai: Zhejiang sheng shuichan shiyanchang, 1936.

———. "Yin xian yuye zhi diaocha" (An investigation of Yin county's fisheries). *Zhejiang sheng jianshe yuekan* 10, no. 4 (1936).

Lin Shuyan and Huang Shubiao. *Zhejiang zhangwang yingxiang yulei fanzhi zhi yanjiu* (An investigation of the effects of stow nets on the reproduction of fish species). Zhejiang sheng shuichan shiyanchang huibao, 3, no. 2. Dinghai: Zhejiang sheng shuichan shiyanchang, 1937.

Ling Lingjiu and Cheng Tiyun. "Zhe Zhao zhuanyuan hua Sheng gui Zhe liyou zhi jiantao ji Jiang-Zhe haijiang tushuo" (A critique of Zhejiang's Commissioner Zhao's reasons for placing Shengsi under Zhejiang jurisdiction and an illustration of the Jiangsu-Zhejiang maritime border). *Jiangsu yanjiu* 1, no. 5 (1935).

Liu, Jianguo, and Jared Diamond. "China's Environment in a Globalizing World: How China and the Rest of the World Affect Each Other." *Nature*, no. 435 (June 30, 2006).

Liu Tongshan and Xu Jibo. "Zhongguo yanhai yuye yu yumin shenghuo" (China's coastal fisheries and the life of fishermen). *Xin Zhonghua* 3, no. 13 (1935).

Liu Zhengquan. "Shengsi liedao jubian" (The Shengsi Islands' great transformation). *Shengsi wenshi ziliao* 2 (1991).

Lu Yanghao. "Cong yuye guandian lun Shengsi de fen'ge" (A discussion of Shengsi's partition from the perspective of fisheries). *Shuichan yuekan* 3, no. 3–4 (1936).

———. "Pi Zhe sheng 'Ningbo,' 'Fenghua,' 'Dinghai' Hu tongxianghui qing hua Shengsi liedao gui Zhe shuo" (A refutation of the request from Zhejiang province's "Ningbo," "Fenghua," and "Dinghai" native-place associations in Shanghai to place the Shengsi Islands under Zhejiang's jurisdiction). *Jiangsu yanjiu* 1, no. 3 (1935).

———. "Su sheng waihai ying fou zhengshou yuye linshi yingye shui zhi shangque" (A discussion of whether Jiangsu province should collect a temporary business tax on fisheries). *Jiangsu yanjiu* 1, no. 4 (1935).

Mann [Jones], Susan. "Finance in Ning-po: The *Ch'ien-chuang*." In *Economic Organization in Chinese Society*, ed. W. E. Willmott. Stanford: Stanford University Press, 1972.

———. *Local Merchants and the Chinese Bureaucracy, 1750–1959*. Stanford: Stanford University Press, 1987.

———. "Women's Work in the Ningbo Area, 1900–1936." In *Chinese History in Economic Perspective*, ed. Thomas R. Rawski and Lillian M. Li. Berkeley: University of California Press, 1992.

Mao Yihu. "Zhuang Songfu de yisheng" (The life of Zhuang Songfu). *Ningbo wenshi ziliao* 4 (1986).

Marks, Robert B. *Tigers, Rice, Silk, and Silt: Environment and Economy in Late Imperial South China*. Cambridge: Cambridge University Press, 1998.

Martin, Brian G. *The Shanghai Green Gang: Politics and Organized Crime, 1919–1937*. Berkeley: University of California Press, 1996.

Matsuda Yoshirō. "Min Shin jidai Sekkō Gin ken no suiri jigyō" (Water control affairs in Zhejiang's Yin county during the Ming and Qing periods). In *Satō hakushi kanreki kinen Chūgoku suiri shi ronsō*, ed. Chūgoku suiri shi kenkyūkai, 269–312. Tokyo: Kokusho kankōkai, 1981.

Matsuura Akira. *Chūgoku no kaizoku* (Chinese pirates). Tokyo: Tōhō shoten, 1995.

Mazumdar, Sucheta. *Sugar and Society in China: Peasants, Technology, and the World Market*. Cambridge: Harvard University Asia Center, 1998.

McEvoy, Arthur F. *The Fisherman's Problem: Ecology and Law in the California Fisheries, 1850–1980*. Cambridge: Cambridge University Press, 1986.

————. "Toward an Interactive Theory of Nature and Culture." In *The Ends of the Earth: Perspectives on Modern Environmental History*, ed. Donald Worster. Cambridge: Cambridge University Press, 1988.

McGoodwin, James. *Crisis in the World's Fisheries: People, Problems, and Politics*. Stanford: Stanford University Press, 1990.

McNeill, J. R. *Something New Under the Sun: An Environmental History of the Twentieth Century World*. New York: Norton, 2000.

Menzies, Nicholas K. *Forest and Land Management in Imperial China*. New York: St. Martin's Press, 1994.

Minguo Xiangshan xianzhi (Republican-period Xiangshan county gazetteer), 1927.

Murray, Dian H. *Pirates of the South China Coast, 1790–1810*. Stanford: Stanford University Press, 1987.

Myers, Ramon H., and Yeh-Chien Wang. "Economic Developments, 1644–1800." In *The Cambridge History of China: The Ch'ing Empire to 1800*, ed. Willard J. Peterson. Cambridge: Cambridge University Press, 2002.

"Nan Shi no suisan" (South China's marine products). In *Nanyō no suisan*. Tokyo: Nanyō suisan kyōkai, 1925.

Naquin, Susan. *Peking: Temples and City Life*. Berkeley: University of California Press, 2000.

Ng, Chin-Keong. *Trade and Society: The Amoy Network on the China Coast, 1683–1735*. Singapore: Singapore University Press, 1983.

Ninohei Tokuo. *Meiji gyogyō kaitaku shi* (History of the expansion of fisheries in the Meiji period). Tokyo: Heibonsha, 1981.

―――. *Nihon gyogyō kindai shi* (The modern history of Japan's fisheries). Tokyo: Heibonsha, 1999.

Okamoto Nobuo. *Kindai gyogyō hattatsu shi* (History of modern fishery development). Tokyo: Suisansha, 1965.

Okamoto Shoichi. *Man-Shi no suisan jijō* (The state of Manchuria and China's marine products). Tokyo: Suisan tsūshinsha, 1940.

Okano Ichirō. "Shina shin seifu no suisan seisaku" (The new government of China's marine products policies). *Suisankai,* no. 563 (1929).

Olson, Mancur. *The Logic of Collective Action: Public Goods and the Theory of Groups.* Cambridge: Harvard University Press, 1965.

Osborne, Anne [Rankin]. "Barren Mountains, Raging Rivers: The Ecological and Social Effects of Changing Landuse on the Lower Yangzi Periphery in Late Imperial China." PhD diss., Columbia University, 1989.

―――. "Highlands and Lowlands: Economic and Ecological Interactions in the Lower Yangzi Region Under the Qing." In *Sediments of Time,* ed. Mark Elvin and Liu Ts'ui-jung. Cambridge: Cambridge University Press, 1998.

Ostrom, Elinor. *Governing the Commons: The Evolution of Institutions for Collective Action.* Cambridge: Cambridge University Press, 1990.

Ostrom, Elinor; Roy Gardner; and James Walker. *Rules, Games, and Common-Pool Resources.* Ann Arbor: University of Michigan Press, 1994.

Ouyang Zongshu. *Haishang renjia: haiyang yuye jingji yu yumin shehui* (Sea people: the marine fishery economy and fishing people's society). Nanchang: Jiangxi gaoxiao chubanshe, 1998.

Perdue, Peter C. *Exhausting the Earth: State and Peasant in Hunan, 1500–1850.* Cambridge: Council on East Asian Studies, Harvard University, 1987.

―――. "Lakes of Empire: Man and Water in Chinese History." *Modern China* 16, no. 1 (1990).

Perry, Elizabeth J. *Rebels and Revolutionaries in North China, 1845–1945.* Stanford: Stanford University Press, 1980.

Pettus, Thomas F. "Cuttle-fish Trade at Ningpo." *Reports from the Consuls of the United States,* vol. 30, nos. 105–7½. Washington, DC: Government Printing Office, 1889.

264

Works Cited

Pietz, David A. *Engineering the State: The Huai River and Reconstruction in Nationalist China, 1927–1937*. New York: Routledge, 2002.

Ping (pseud.). "Shatian xian ken chuyan" (My humble remarks on the reclamation of polder fields by the county). *Jiangsu yanjiu* 1, no. 7 (1935).

Prattis, J. I. "Modernization and Modes of Production in the North Atlantic: A Critique of Policy Formation for the Development of Marginal Maritime Communities." *American Journal of Economics and Sociology* 39, no. 4 (1980).

Qin Yunshan et al. *Geology of the East China Sea*. Beijing: Science Press, 1996.

Qiu Yongsong et al. "Runoff- and Monsoon-Driven Variability of Fish Production in East China Seas." *Estuarine, Coastal and Shelf Science* 77 (2008).

Qiu Zhonglin. "Bingxian chuan yu xianyu hang: Ming dai yijiang Jiangnan de bingxian yuye yu haixian xiaofei" (Ice boats and fresh fish brokers: the frozen fish industry and seafood consumption in Jiangnan since the Ming period). In *"Guoyan fanhua: Ming-Qing Jiangnan de shenghuo yu wenhua" guoji xueshu taolunhui*. Nangang, Taiwan: Zhongyang yanjiuyuan, Lishi yuyan yanjiusuo, 2003.

Qu Ruoquan. "Jiang Zhe zhengyi zhong zhi Shengsi huazhi wenti" (The problem of Shengsi's jurisdiction in the dispute between Jiangsu and Zhejiang). *Shuichan yuekan* 3, no. 3/4 (1936).

Rankin, Mary Backus. *Elite Activism and Political Transformation in China: Zhejiang Province, 1865–1911*. Stanford: Stanford University Press, 1986.

Rawski, Thomas. *Economic Growth in Prewar China*. Berkeley: University of California Press, 1989.

Read, Bernard E. *Common Food Fishes of Shanghai*. Shanghai: North China Branch of the Royal Asiatic Society, 1939.

Reeves, Jesse S. "The Codification of the Law of Territorial Waters." *American Journal of International Law* 24, no. 3 (1930).

"Report of the Second Committee." *American Journal of International Law* 24, no. 3, Supplement: Official Documents (1930).

Rowe, William T. *Crimson Rain: Seven Centuries of Violence in a Chinese County*. Stanford: Stanford University Press, 2007.

———. *Hankow: Commerce and Society in a Chinese City, 1796–1889*. Stanford: Stanford University Press, 1984.

Scheiber, Harry N. *Inter-Allied Conflicts and Ocean Law, 1945–1953: The Occupation Command's Revival of Japanese Whaling and Marine Fisheries.* Taipei: Institute of European and American Studies, Academia Sinica, 2001.

Schlager, Edella. "Fishers' Institutional Reponses to Common-Pool Resource Dilemmas." In *Rules, Games, and Common-Pool Resources,* ed. Elinor Ostrom, Roy Gardner, and James Walker. Ann Arbor: University of Michigan Press, 1994.

Schneider, Laurence. *Biology and Revolution in Twentieth-Century China.* Lanham, MD: Rowan and Littlefield, 2003.

Schoppa, R. Keith. *Song Full of Tears: Nine Centuries of Chinese Life at Xiang Lake.* Boulder, CO: Westview, 2002.

Scoones, Ian. "Range Management Science and Policy: Politics, Polemics and Pasture in Southern Africa." In *The Lie of the Land: Challenging Received Wisdom on the African Environment,* ed. Melissa Leach and Robin Mearns. London: International African Studies Institute, 1996.

Scott, James C. *Seeing Like a State: How Certain Schemes to Improve the Human Condition Have Failed.* New Haven: Yale University Press, 1998.

Shang Hong. "Haiyang yuye ziyuan baohu yu guanli wenti de tantao" (An inquiry into the problem of the protection and management of marine fishery resources). In *Zhoushan haiyang yu yuye jingji yanjiu wenxuan,* ed. Cai Hongzhou. Beijing: Beijing wenjin chubanshe, 2003.

Shanghai bowuguan. Tushu ziliao shi. *Shanghai beike ziliao xuanji* (Collection of Shanghai inscriptional materials). Shanghai: Shanghai renmin chubanshe, 1980.

Shanghai shi nianjian (Shanghai municipal yearbook). Shanghai: Shanghai shi tongzhiguan, 1937.

Shanghai yuyezhi bianzuan weiyuanhui. *Shanghai yuyezhi* (Shanghai fishery gazetteer). Shanghai: Shanghai shehui kexue yanjiuyuan, 1998.

Shao, Qin. *Culturing Modernity: The Nantong Model, 1890–1930.* Stanford: Stanford University Press, 2004.

Shapiro, Judith. *Mao's War Against Nature: Politics and Environment in Revolutionary China.* Cambridge: Cambridge University Press, 2001.

Shen Guangshi. "Jiang Zhe yuye shicha baogao" (Inspection report on Jiangsu's and Zhejiang's fisheries). *Nongkuang gongbao* 8 (1929).

Shen Ligong. "Qiantan Gaoting de liuwang zuoye" (Elementary introduction to Gaoting's drift-net fishing). *Daishan wenshi ziliao* 1 (1986).

Sheng Guanxi. "Jindai Zhoushan de diandangye" (Modern Zhoushan's pawnshops). *Zhoushan xiangxun* 60 (1996).

Shenjiamen zhenzhi bianzuan lingdao xiaozu. *Shenjiamen zhenzhi* (Shenjiamen township gazetteer). Hangzhou: Zhejiang renmin chubanshe, 1996.

Shiba, Yoshinobu. "Environment Versus Water Control: The Case of the Southern Hangzhou Bay Area from the Mid-Tang Through the Qing." In *Sediments of Time*, ed. Mark Elvin and Liu Ts'ui-jung. Cambridge: Cambridge University Press, 1998.

———. "Ningpo and Its Hinterland." In *The City in Late Imperial China*, ed. G. William Skinner. Stanford: Stanford University Press, 1977.

Shindo Shigeaki. "A Statistical Account of the Japanese Trawl Fishery in the East China and the Yellow Seas After the War II [*sic*]." Text in Japanese. *The Investigations of Demersal Fish Resources in the East China and Yellow Seas* 3 (1956).

Simoons, Frederick, J. *Food in China: A Cultural and Historical Inquiry*. Boca Raton, FL: CRC Press, 1991.

Skinner, G. William. "Introduction." In Sow-Theng Leong, *Migration and Ethnicity in Chinese History: Hakkas, Pengmin, and Their Neighbors*, ed. Tim Wright, with an introduction and maps by G. William Skinner. Stanford: Stanford University Press, 1997.

———. "Regional Urbanization in Nineteenth-Century China." In *The City in Late Imperial China*, ed. idem. Stanford: Stanford University Press, 1977.

Smil, Vaclav. *China's Environmental Crisis: An Inquiry into the Limits of National Development*. Armonk, NY: M. E. Sharpe, 1993.

———. *China's Past, China's Future: Energy, Food, Environment*. New York: Routledge, 2004.

Smith, Tim D. *Scaling Fisheries: The Science of Measuring the Effects of Fishing, 1855–1955*. Cambridge: Cambridge University Press, 1994.

Solecki, Jan J. *Economic Aspects of the Fishing Industry in Mainland China*. Vancouver: Institute of Fisheries, University of British Columbia, 1966.

Song, Yann-huei Billy. "China's Ocean Policy: EEZ and Marine Fisheries." *Asian Survey* 29, no. 10 (1989).

Songster, E. Elena. "Cultivating the Nation in Fujian's Forests: Forest Policies and Afforestation Efforts in China, 1911–1937." *Environmental History* 8, no. 3 (2003).

Steinberg, Ted. "Down to Earth: Nature, Agency, and Power in History." *American Historical Review* 107, no. 3 (2002).

Steneck, Robert S., and James T. Carlton. "Human Alterations of Marine Communities: Students Beware!" In *Marine Community Ecology*, ed. Steven D. Gaines, Mark D. Bertness, and Mark E. Hay. Sunderland, MA: Sinauer Associates, 2001.

Sun Biaoqing and Mao Yihu. "Xinxue huishe ji qita" (The New Learning Company and other matters). *Ningbo wenshi ziliao* 7 (1987).

Sun Yat-sen. *The International Development of China*. Taibei: China Cultural Service, 1953.

Tachi Sakutarō. *Heiji kokusaihō ron* (Theory of peacetime international law). Tokyo: Nihon hyōronsha, 1930.

Taizhou fuzhi (Taizhou prefectural gazetteer). 1936.

Takenobu, Y. *The Japan Year Book: Complete Cyclopaedia of General Information and Statistics on Japan and Japanese Territories for the Year 1928*. Tokyo: Japan Year Book Office, 1928.

Takumukyoku. *Chūnan Shina hōmen ni okeru suisan jijō* (The state of marine products in southeast China). Tokyo: Takumushō Takumukyoku, 1938.

Tao Fusheng. "Jiazhi de yuye" (Jiazhi's fisheries). *Jiaojiang wenshi ziliao* 7 (1989).

Tōa dōbunkai. *Shina shōbetsu zenshi*, vol. 14, *Sekkō shō* (Complete gazetteer of China's provinces: Zhejiang province). Tokyo, 1917–20.

Tsutsui, William. "Landscapes in the Dark Valley: Toward an Environmental History of Wartime Japan." *Environmental History* 8, no. 2 (2003).

Tu Hengting. "Minguo shiqi Daishan jinrongye gaikuang" (Finance in Daishan during the Republican period). *Daishan wenshi ziliao* 4 (1992).

Tung, L. *China and Some Phases of International Law*. Shanghai: Kelly and Walsh, 1940.

Waldron, Arthur. *From War to Nationalism: China's Turning Point, 1924–1925*. Cambridge: Cambridge University Press, 1995.

Walker, Brett L. "Meiji Modernization, Scientific Agriculture, and the Destruction of Japan's Hokkaido Wolf." *Environmental History* 9, no. 2 (2004).

Wang, Bin. *The Asian Monsoon*. London: Springer, 2006.

Wang Ningshi. *Zhejiang yanhai ge xian yu yan gaikuang* (The general condition of fisheries and salt in Zhejiang's coastal provinces). Ningbo: Zhejiang shengli Ningbo minzhong jiaoyuguan, 1936.

Wang Renze. "Lü Hu mingren Fang Jiaobo" (The famous Shanghai so-journer Fang Jiaobo). *Ningbo wenshi ziliao* 5 (1987).

Wang Rongguo. *Haiyang shenling: Zhongguo haishen xinyang yu shehui jingji* (Ocean gods: China's maritime deity beliefs and social economy). Nanchang: Jiangxi gaoxiao chubanshe, 2003.

Wang Weimin. "Huiyi Zhuang Songfu xiansheng" (Rembering Mr. Zhuang Songfu). *Fenghua wenshi ziliao* 3 (1987).

Wang Zenglu. "Jiangsu waihai caiwu xingzheng diaocha" (Investigation of offshore fiscal administration in Jiangsu). *Jiangsu yanjiu* 1, no. 8 (1935).

Wang Zongpei. "Zhongguo yanhai zhi yumin jingji" (Fishermen's economy in coastal China). *Jingjixue jikan* 3, no. 1 (1932).

Ward, Barbara E. "Chinese Fishermen in Hong Kong: Their Post-Peasant Economy." In *Social Organization: Essays Presented to Raymond Firth*, ed. Maurice Freedman. Chicago: Aldine, 1967.

Watson, James L. "Standardizing the Gods: The Promotion of T'ien Hou ("Empress of Heaven") Along the South China Coast, 960–1960." In *Popular Culture in Late Imperial China*, ed. David Johnson, Andrew Nathan, and Evelyn Rawski. Berkeley: University of California Press, 1985.

Weller, Robert P. *Discovering Nature: Globalization and Environmental Culture in China and Taiwan*. Cambridge: Cambridge University Press, 2006.

Weng Yingchang. "Tangtou yumin he Mazu wenhua" (Tangtou fisher-men and Mazu culture). In *Wenshi tiandi*, ed. Zhoushan shi zhengxie wenshi he xuexi wei. Beijing: Wenjin chubanshe, 2003.

Wenling xian xuzhi gao bianzuan weiyuanhui. *Wenling xian xuzhi gao* (Wenling county extended draft gazetteer). Taibei: Taibei shi Wen-ling tongxianghui, 1987.

Williamsen, Marvin. "The Military Dimension, 1937–1941." In *China's Bitter Victory: The War With Japan 1937–1945*, ed. James C. Hsiung and Steven I. Levine. Armonk, NY: M. E. Sharpe, 1992.

Worcester, G. R. G. *The Floating Population in China: An Illustrated Record of the Junkmen and Their Boats on Sea and River*. Hong Kong: Vetch and Lee, 1970.

———. *The Junks and Sampans of the Yangtze. A Study in Chinese Nautical Research*. 2 vols. Shanghai: Statistical Department of the Inspectorate General of Customs, 1947.

Worm, Boris, et al. "The Impacts of Biodiversity Loss on Ocean Eco-system Services." *Science* 313 (November 3, 2006).

Wu Xingya. *Minguo ershiyi nian Shanghai shi yulun huigu* (A review of Shanghai's fishing trawlers in 1932). Shanghai, 1934.

Wu Zaisheng. "Shehui shenghuo: Fenghua de buyu jia" (Social life: Feng-hua's fishing households). *Xin Zhongguo* 1, no. 6 (1919).

Xie Zhenmin. *Zhonghua minguo lifa shi* (Legislative history of Republican China). Beijing: Zhongguo zhengfa daxue chubanshe, 1999.

Xin xiu Yin xianzhi (Revised Yin county gazetteer). 1877.

Xu Bin. "Penglai hangdu shu Nanpu" (Nanpu's Penglai ferry). *Daishan wenshi ziliao* 1 (1986).

Xu Bo. *Zhoushan fangyan yu Donghai wenhua* (Zhoushan dialect and East China Sea culture). Beijing: Zhongguo shehui kexue chubanshe, 2004.

Xu Shihe. "Jiang Zhe liang sheng zhi yuye" (The fisheries of Jiangsu and Zhejiang). *Shenbao*, October 6, 1925, 1 (supplement).

Yang, S. K., et al. "Trends in Annual Discharge from the Yangtze River to the Sea (1865–2004)." *Hydrological Sciences* 50, no. 5 (2005).

Yao Huanzhou. "Zhoushan qundao wuzei zhi shengxi ji wangbu yu longbu zhi deshi" (The reproduction of cuttlefish in the Zhoushan Archipelago and the advantages and disadvantages of net fishing and cage fishing). In *Zhejiang sheng zhengfu jiansheting ershiyi nian niankan*. Hangzhou, 1932.

Yao Yongping. "Daishan shuichan zhi diaocha" (An investigation of Daishan's marine products). *Zhejiang sheng jianshe yuekan* 6, no. 7 (1933).

———. "Gaijin Zhejiang dahuangyu yuye ji zhizaoye zhi yijian" (Ideas for the reform of Zhejiang's large yellow croaker fisheries and pro-cessing industry). *Zhejiang sheng jianshe yuekan* 7, no. 9 (1934).

Yi Lingling. *Ming Qing Changjiang zhong xia you yuye jingji yanjiu* (Re-searches on the fishery economy of the Middle and Lower Yangzi during the Ming and Qing). Ji'nan: Jilu shushe, 2004.

Yin xian tongzhi (Yin county gazetteer). 1935.

Ying Mengqing. "Fenghua yumin canjia guangfu Hangzhou gansidui ji" (A record of Fenghua fishermen's participation in a dare-to-die squad during the recovery of Hangzhou). In *Zhejiang xinhai geming huiyi lu*. Hangzhou: Zhejiang renmin chubanshe, 1981.

———. "Fenghua yumin gansidui canjia guangfu Hangzhou" (The par-ticipation of Fenghua fishermen's dare-to-die squad in the recovery of Hangzhou). *Fenghua wenshi ziliao* 11 (1991).

Young, Oran. *The Institutional Dimensions of Environmental Change: Fit, Interplay, and Scale*. Cambridge: MIT Press, 2002.

Yu Fuhai, ed. *Ningbo shizhi waibian* (Ningbo municipal gazetteer, additional section). Beijing: Zhonghua shuju, 1998.

Yu Men. "Shenjiamen Tian Hou gong" (Shenjiamen's Empress of Heaven Temple). In *Wenshi tiandi*, ed. Zhoushan shi zhengxie wenshi he xuexi wei. Beijing: Wenjin chubanshe, 2003.

Zhang Baoshu. *Zhongguo yuye* (China's fisheries). 2 vols. Taibei: Zhongguo wenhua chuban shiye weiyuanhui, 1954.

Zhang Genfu. *Kangzhan shiqi Zhejiang sheng renkou qianyi yu shehui yingxiang* (Population mobility in Zhejiang province during the War of Resistance and its social effects). Shanghai: Shanghai sanlian shudian, 2001.

Zhang Jiacheng and Lin Zhiguang. *Climate of China*. Trans. Ding Tan. New York: John Wiley and Sons, 1992.

Zhang Jian. "Yumin 'xie yang'" (Fishermen "thanking the ocean"). In *Zhoushan minsu daguan*, ed. idem. Beijing: Yuanfang chubanshe, 1999.

Zhang Liquan. "Shanghai haichanwu shichang zhi qushi" (Trends in Shanghai's marine products market). *Shuichan* 2 (1918).

Zhang Lixiu and Bi Dingbang. *Zhejiang dangdai yuye shi* (History of Zhejiang's contemporary fisheries). Hangzhou: Zhejiang kexue jishu chubanshe, 1990.

Zhang Qiang et al. "Observed Climatic Changes in Shanghai During 1873–2002." *Journal of Geographical Sciences* 15, no. 2 (2005).

Zhang Qijun. *Zhejiang sheng shidi jiyao* (Historical geography of Zhejiang). Shanghai: Shangwu yinshuguan, 1925.

Zhang Qilong and Wang Fan. "Zhoushan yuchang ji qi linjin haiyu shuituan de qihouxue fenxi" (Climatological analysis of Zhoushan's fishing grounds and water masses in nearby seas). *Haiyang yu huzhao* 35, no. 1 (2004).

Zhang Renyu. "Xinhai geming yilai de Dinghai shangye yanbian jianshi" (Brief history of the development of commerce in Dinghai since the 1911 Revolution). *Zhoushan wenshi ziliao* 1 (1991).

Zhang Zhendong and Yang Jinsen. *Zhongguo haiyang yuye jianshi* (Brief history of China's marine fisheries). Beijing: Haiyang chubanshe, 1983.

Zhang Zhuzun. "Fazhan Zhejiang shuichan jiaoyu banfa yijian shu" (Opinions on methods for the development of Zhejiang's marine products education). *Zhejiang shengli shuichanke zhiye xuexiao xiaokan* (1929).

Zhao Chuanyin et al. "Changjiang jingliu dui hekou ji linjin haiqu yuye yingxiang de chubu yanjiu" (Preliminary studies on the influence of Changjiang river runoff on the fisheries of Changjiang estuary and its adjacent seas). *Shuichan xuebao* 12, no. 4 (1988).

Zhao Yizhong. "Zhonghua renmin gongheguo jianguo chuqi huifu Zhoushan yuye shengchan de ji ge licheng" (Courses to the recovery of fishery production in Zhoushan during the early period of the People's Republic of China). *Dinghai wenshi ziliao* 1 (1989).

————. "Zhoushan de bingxian shang he yuhang jianxi" (An analysis of Zhoushan's frozen fish merchants and fish brokers). *Dinghai wenshi ziliao* 2 (1985).

————. "Zhoushan de yuye gongsuo" (Zhoushan's fishing lodges). *Zhoushan wenshi ziliao* 1 (1991).

————. "Zhoushan yuye fazhan shi chutan" (A preliminary investigation of the history of the development of Zhoushan's fisheries). *Zhongguo shehui jingji shi yanjiu* 9, no. 2 (1984).

Zhejiang sheng. Jiansheting. *Liang nian lai zhi Zhejiang jianshe gaikuang* (The general condition of reconstruction in Zhejiang over the past two years). Hangzhou, 1929.

————. Shuichanting. *Lüsi yuchang shengchan ziliao (wenjian): Zhejiang yumin zai Lüsi yuchang shengchan* (Materials on fishery production in the Lüsi fishing grounds (documents): Zhejiang fishermen's production in the Lüsi fishing grounds), 1967–1971.

————. Shuiliju. *Zhejiang sheng shuiliju zongbaogao* (General report of the Water Control Bureau of Zhejiang province). Hangzhou: 1935.

Zhejiang sheng ge xian jianshe gongzuo taolunhui mishuchu. *Zhejiang sheng ge xian jianshe gongzuo taolunhui huiyi baogao* (Reports from the conference to discuss reconstruction work in each of the counties of Zhejiang province). 3 vols. 1939.

Zhejiang sheng haian he haitu ziyuan zonghe diaocha baogao bianxie weiyuanhui. *Zhejiang sheng haian he haitu ziyuan zonghe diaocha* (A comprehensive investigation of Zhejiang province's coastal and tidal-estuary resources). Beijing: Haiyang chubanshe, 1988.

Zhejiang shengli shuichan shiyanchang. "Shengshan moyu fanzhi shiyan baogao" (Shengshan cuttlefish-spawning experiment report). *Shuichan yuekan* 3, no. 5–6 (1936).

Zhejiang sheng shuichanzhi bianzuan weiyuanhui. *Zhejiang sheng shuichanzhi* (Zhejiang province marine products gazetteer). Beijing: Zhonghua shuju, 1999.

Zhejiang sheng Yin xianzhi bianzuan weiyuanhui. *Yin xianzhi* (Yin county gazetteer). Beijing: Zhonghua shuju, 1996.

Zhejiang sheng zhengxie wenshi zilao weiyuanhui. *Zhejiang jinxiandai renwu lu* (Record of modern and contemporary personages in Zhejiang). Shangyu: Zhejiang renmin chubanshe, 1992.

Zheng Pingsheng. "Haiyang ziyuan guanli cuoshi chutan" (A preliminary investigation of measures for the management of marine resources). In *Zhoushan haiyang yu yuye jingji yanjiu wenxuan*, ed. Cai Hongzhou. Beijing: Beijing wenjin chubanshe, 2003.

Zheng Ruozeng. *Chouhai tubian* (Illustrated collection on coastal defense). 1562.

Zhenhai xianzhi (Zhenhai county gazetteer). 1752.

Zhenhai xianzhi bianzuan weiyuanhui. *Zhenhai xianzhi* (Zhenhai county gazetteer). Shanghai: Zhongguo da baikequanshu chubanshe, 1994.

Zhongguo dier lishi dang'anguan. *Zhonghua minguo shi dang'an ziliao huibian* (Compilation of archival materials on the history of the Republic of China). Nanjing: Jiangsu guji chubanshe, 1991.

Zhongguo nongye baikequanshu (China agriculture encyclopedia), vols. 11–12, *Shuichan* (Marine products). Beijing: Nongye chubanshe, 1991.

Zhongguo shiye zhi, Zhejiang sheng (Record of China's industries, Zhejiang province). Shanghai: Shiyebu, Guoji maoyiju, 1933.

Zhong-Ri guanxi shiliao: yu yan lu kuang jiaoshe minguo qi nian zhi shiliu nian (1918–27) (Materials on Sino-Japanese relations: fisheries, salt, railroad, and mining negotiations, 1918–1927). Taibei: Zhongyang yanjiuyuan, Jindaishi yanjiusuo, 1995.

Zhou Jianyin and Yu Huaxian. *Zhongdeng shuichan xue* (Intermediate fishery studies). Shanghai: Zhonghua shuju, 1928.

Zhou Kebao and Lu Yonglong. "Ershi nian Zhejiang sheng zhi shuizai" (Flooding in Zhejiang province during 1931). *Zhejiang sheng jianshe yuekan* 6, no. 11 (1933).

"Zhoushan qundao zhi yuchang" (The Zhoushan Archipelago's fishing grounds). *Yinhang zhoubao* 17, no. 5 (1933).

Zhoushan shi bowuguan. *Zhoushan lishi mingren pu* (Register of famous historical personages from Zhoushan). Beijing: Zhongguo wenshi chubanshe, 2004.

Zhoushan shi dang'anguan. *"Shenbao" Zhoushan shiliao huibian* (Compilation of Zhoushan historical materials from *Shenbao*). Zhoushan: Zhoushan ribaoshe, 1990.

Zhoushan shi difangzhi bianzuan weiyuanhui. *Zhoushan shizhi* (Zhoushan municipal gazetteer). Hangzhou: Zhejiang renmin chubanshe, 1992.

Zhoushan shizheng wenshi he xuexi weiyuanhui. *Zhoushan haiyang long wenhua* (Zhoushan's maritime dragon culture). Beijing: Haiyang chubanshe, 1999.

———. *Zhoushan haiyang yu wenhua* (Zhoushan's maritime fish culture). Beijing: Haiyang chubanshe, 1992.

Zhoushan yuzhi bianxie zu. *Zhoushan yuzhi* (Zhoushan fishery gazetteer). Beijing: Haiyang chubanshe, 1989.

Zhu Fucheng. *Jiangsu shatian zhi yanjiu* (Researches on polder fields in Jiangsu). Taibei: Chengwen, 1977.

Zhu Jin'gou. "Qingmo he minguo shi Huanglong dao shang shou hang chuan" (Net-protection boats on Huanglong Island during the late Qing and Republican periods). *Shengsi wenshi ziliao* 2 (1991).

Zhu Yunshui. "Zhejiang yutuan yan'ge shi" (History of the development of the Zhejiang fishing militia). *Zhejiang shengli shuichanke zhiye xuexiao xiaokan* 1929.

Zhu Zhengyuan. *Zhejiang sheng yanhai tushuo* (Illustrated handbook of coastal Zhejiang province). 1899. Reprinted—Taibei: Chengwen, 1974.

Zhuang Jingzhong (Songfu). *Qiuwo shanren nianpu* (Chronicle of the life of Zhuang Songfu). 1929.

Index

275

Shenjiamen Pair Fishing Same-
Trade Association, 153, 161
Shenjiamen Self-Government
Association, 177
Shenjiamen Sojourners Lodge,
153
Sheshan, 102, 114, 122, 164
Shi Meiheng, 161–65 *passim*
Shi Renhang, 92
Shipu, 92
Shishunxing fish brokerage, 54
Short-term business tax, 152–54,
167, 240
Shrimp, 38, 186
Sijiao, 30, 102, 128, 155, 220
Silt, 22, 137, 208
Sino-French War, 49
Sino-Japanese War (1894–95),
49
Sino-Japanese War (1937–45),
148, 174–78
Silver, 33, 59, 78, 111
Sino-French Fishery Company,
175–76
Skinner, G. William, 22, 208
Small pair boats, 23
Small yellow croaker, 16, 17, 102,
185, 186
Smil, Vaclav, 228
Smith, Tim D., 187
Smuggling, 20, 116–20 *passim*,
212, 230
Song dynasty, 17
Song Ziwen, 120
Songjiang, 91
Songmen, 91, 226
South China Sea, 123
Sovereignty, 76, 106, 107, 108,
114–15, 117, 125
Spawning, 16–17, 68, 81, 99–100;
protection of, 82, 84, 111, 184,
186; cuttlefish, 128, 132, 135,

138–41 *passim*, 144–48 *passim*,
155
Special administrative supervisors,
158, 168, 245
Squid, 139
State-building, 8, 193
Steamships, 68, 76–78, 102, 111–12,
118–19, 156, 160–61, 176
Steam trawlers, 76–78, 84–85
Stow nets, 30, 42–43, 53–54, 99,
122, 129, 130, 133
Sturgeon, 106
Sun Biaoqing, 217
Sun Chuanfang, 111–12
Sun Guansheng, 56, 57
Sun Yat-sen, 86, 97, 113, 142, 160,
182

Taihe Lodge, 46, 72
Taiping Rebellion, 21, 22, 53
Taiwan, 103
Taiwan Warm Current, 16
Taizhou, 26, 40, 100, 137, 156,
168, 176, 225; migration from,
21, 22, 23, 28, 67, 92, 165, 214;
piracy, 49, 50, 51; lodges, 52,
78, 91, 92, 155; cage fishing,
132–36, 140, 141, 143, 148, 191,
236
Taizhou Fish Merchants Lodge,
140
Taizhou Lodge, 78, 92
Taizhou Prefectural Commissioners Office, 184–85
Tanxu, 143, 169
Tariffs, 114, 120, 234
Taxes, 131, 137, 171, 172, 184, 212,
225; and lodges, 36, 49, 50, 62,
78, 151; on fisheries, 8, 9, 97,
109–12, 126, 178–79, 192; reduction of, 113, 116, 118, 120; in
Shengsi Islands, 152–53, 154, 157,

Harvard East Asian Monographs
(*out-of-print)

Harvard East Asian Monographs

Harvard East Asian Monographs